197
Advances in Polymer Science

Editorial Board:
A. Abe · A.-C. Albertsson · R. Duncan · K. Dušek · W. H. de Jeu
J.-F. Joanny · H.-H. Kausch · S. Kobayashi · K.-S. Lee · L. Leibler
T. E. Long · I. Manners · M. Möller · O. Nuyken · E. M. Terentjev
B. Voit · G. Wegner · U. Wiesner

Advances in Polymer Science
Recently Published and Forthcoming Volumes

Peptide Hybrid Polymers
Volume Editors: Klok, H.-A., Schlaad, H.
Vol. 202, 2006

**Supramolecular Polymers/
Polymeric Betains**
Vol. 201, 2006

Ordered Polymeric Nanostructures at Surfaces
Volume Editor: Vansco, G. J.
Vol. 200, 2006

Emissive Materials/Nanomaterials
Vol. 199, 2006

Surface-Initiated Polymerization II
Volume Editor: Jordan, R.
Vol. 198, 2006

Surface-Initiated Polymerization I
Volume Editor: Jordan, R.
Vol. 197, 2006

**Conformation-Dependent Design of Sequences
in Copolymers II**
Volume Editor: Khokhlov, A. R.
Vol. 196, 2006

**Conformation-Dependent Design of Sequences
in Copolymers I**
Volume Editor: Khokhlov, A. R.
Vol. 195, 2006

Enzyme-Catalyzed Synthesis of Polymers
Volume Editors: Kobayashi, S., Ritter, H.,
Kaplan, D.
Vol. 194, 2006

Polymer Therapeutics II
Polymers as Drugs, Conjugates and Gene
Delivery Systems
Volume Editors: Satchi-Fainaro, R., Duncan, R.
Vol. 193, 2006

Polymer Therapeutics I
Polymers as Drugs, Conjugates and Gene
Delivery Systems
Volume Editors: Satchi-Fainaro, R., Duncan, R.
Vol. 192, 2006

**Interphases and Mesophases in Polymer
Crystallization III**
Volume Editor: Allegra, G.
Vol. 191, 2005

Block Copolymers II
Volume Editor: Abetz, V.
Vol. 190, 2005

Block Copolymers I
Volume Editor: Abetz, V.
Vol. 189, 2005

**Intrinsic Molecular Mobility and Toughness
of Polymers II**
Volume Editor: Kausch, H.-H.
Vol. 188, 2005

**Intrinsic Molecular Mobility and Toughness
of Polymers I**
Volume Editor: Kausch, H.-H.
Vol. 187, 2005

Polysaccharides I
Structure, Characterization and Use
Volume Editor: Heinze, T.
Vol. 186, 2005

**Advanced Computer Simulation
Approaches for Soft Matter Sciences II**
Volume Editors: Holm, C., Kremer, K.
Vol. 185, 2005

Crosslinking in Materials Science
Vol. 184, 2005

Surface-Initiated Polymerization I

Volume Editor: Rainer Jordan

With contributions by
R. Advincula · M. R. Buchmeiser · D. J. Dyer · T. Fukuda
A. Goto · T. Matsuda · K. Ohno · Y. Tsujii · S. Yamamoto

The series *Advances in Polymer Science* presents critical reviews of the present and future trends in polymer and biopolymer science including chemistry, physical chemistry, physics and material science. It is adressed to all scientists at universities and in industry who wish to keep abreast of advances in the topics covered.
As a rule, contributions are specially commissioned. The editors and publishers will, however, always be pleased to receive suggestions and supplementary information. Papers are accepted for *Advances in Polymer Science* in English.
In references *Advances in Polymer Science* is abbreviated *Adv Polym Sci* and is cited as a journal.

Springer WWW home page: http://www.springer.com
Visit the APS content at http://www.springerlink.com/

Library of Congress Control Number: 2005935448

ISSN 0065-3195
ISBN-10 3-540-30247-6 Springer Berlin Heidelberg New York
ISBN-13 978-3-540-30247-6 Springer Berlin Heidelberg New York
DOI 10.1007/11569084

This work is subject to copyright. All rights are reserved, whether the whole or part of the material is concerned, specifically the rights of translation, reprinting, reuse of illustrations, recitation, broadcasting, reproduction on microfilm or in any other way, and storage in data banks. Duplication of this publication or parts thereof is permitted only under the provisions of the German Copyright Law of September 9, 1965, in its current version, and permission for use must always be obtained from Springer. Violations are liable for prosecution under the German Copyright Law.

Springer is a part of Springer Science+Business Media

springer.com

© Springer-Verlag Berlin Heidelberg 2006
Printed in Germany

The use of registered names, trademarks, etc. in this publication does not imply, even in the absence of a specific statement, that such names are exempt from the relevant protective laws and regulations and therefore free for general use.

Cover design: *Design & Production* GmbH, Heidelberg
Typesetting and Production: LE-TEX Jelonek, Schmidt & Vöckler GbR, Leipzig

Printed on acid-free paper   02/3100 YL – 5 4 3 2 1 0

Volume Editor

Dr. Rainer Jordan
Lehrstuhl für Makromolekulare Stoffe
Lichtenbergstr. 4
TU München
85747 Garching, Germany
Rainer.Jordan@ch.tum.de

Editorial Board

Prof. Akihiro Abe
Department of Industrial Chemistry
Tokyo Institute of Polytechnics
1583 Iiyama, Atsugi-shi 243-02, Japan
aabe@chem.t-kougei.ac.jp

Prof. A.-C. Albertsson
Department of Polymer Technology
The Royal Institute of Technology
10044 Stockholm, Sweden
aila@polymer.kth.se

Prof. Ruth Duncan
Welsh School of Pharmacy
Cardiff University
Redwood Building
King Edward VII Avenue
Cardiff CF 10 3XF, UK
DuncanR@cf.ac.uk

Prof. Karel Dušek
Institute of Macromolecular Chemistry, Czech
Academy of Sciences of the Czech Republic
Heyrovský Sq. 2
16206 Prague 6, Czech Republic
dusek@imc.cas.cz

Prof. W. H. de Jeu
FOM-Institute AMOLF
Kruislaan 407
1098 SJ Amsterdam, The Netherlands
dejeu@amolf.nl
and Dutch Polymer Institute
Eindhoven University of Technology
PO Box 513
5600 MB Eindhoven, The Netherlands

Prof. Jean-François Joanny
Physicochimie Curie
Institut Curie section recherche
26 rue d'Ulm
75248 Paris cedex 05, France
jean-francois.joanny@curie.fr

Prof. Hans-Henning Kausch
Ecole Polytechnique Fédérale de Lausanne
Science de Base
Station 6
1015 Lausanne, Switzerland
kausch.cully@bluewin.ch

Prof. Shiro Kobayashi
R & D Center for Bio-based Materials
Kyoto Institute of Technology
Matsugasaki, Sakyo-ku
Kyoto 606-8585, Japan
kobayash@kit.ac.jp

Prof. Kwang-Sup Lee
Department of Polymer Science &
Engineering
Hannam University
133 Ojung-Dong
Daejeon 306-791, Korea
kslee@hannam.ac.kr

Prof. L. Leibler
Matière Molle et Chimie
Ecole Supérieure de Physique
et Chimie Industrielles (ESPCI)
10 rue Vauquelin
75231 Paris Cedex 05, France
ludwik.leibler@espci.fr

Prof. Timothy E. Long
Department of Chemistry
and Research Institute
Virginia Tech
2110 Hahn Hall (0344)
Blacksburg, VA 24061, USA
telong@vt.edu

Prof. Ian Manners
School of Chemistry
University of Bristol
Cantock's Close
BS8 1TS Bristol, UK
ian.manners@bristol.ac.uk

Prof. Martin Möller
Deutsches Wollforschungsinstitut
an der RWTH Aachen e.V.
Pauwelsstraße 8
52056 Aachen, Germany
moeller@dwi.rwth-aachen.de

Prof. Oskar Nuyken
Lehrstuhl für Makromolekulare Stoffe
TU München
Lichtenbergstr. 4
85747 Garching, Germany
oskar.nuyken@ch.tum.de

Prof. E. M. Terentjev
Cavendish Laboratory
Madingley Road
Cambridge CB 3 OHE, UK
emt1000@cam.ac.uk

Prof. Brigitte Voit
Institut für Polymerforschung Dresden
Hohe Straße 6
01069 Dresden, Germany
voit@ipfdd.de

Prof. Gerhard Wegner
Max-Planck-Institut
für Polymerforschung
Ackermannweg 10
Postfach 3148
55128 Mainz, Germany
wegner@mpip-mainz.mpg.de

Prof. Ulrich Wiesner
Materials Science & Engineering
Cornell University
329 Bard Hall
Ithaca, NY 14853, USA
ubw1@cornell.edu

Advances in Polymer Science
Also Available Electronically

For all customers who have a standing order to Advances in Polymer Science, we offer the electronic version via SpringerLink free of charge. Please contact your librarian who can receive a password or free access to the full articles by registering at:

springerlink.com

If you do not have a subscription, you can still view the tables of contents of the volumes and the abstract of each article by going to the SpringerLink Homepage, clicking on "Browse by Online Libraries", then "Chemical Sciences", and finally choose Advances in Polymer Science.

You will find information about the

- Editorial Board
- Aims and Scope
- Instructions for Authors
- Sample Contribution

at springer.com using the search function.

Preface

These two volumes on surface-initiated polymerization deal with recent developments in the synthesis, characterization and properties of structurally and chemically defined polymer coatings on surfaces. Nearly all polymerization techniques that have been developed in solution have now been adapted for the surface-initiated polymerization (SIP). The reader will find all relevant techniques discussed in these volumes, such as free, controlled and living radical polymerization, living anionic and cationic polymerization (Rigoberto Advincula), and ring-opening metathesis polymerization (Michael Buchmeiser). Most of them are used to prepare so-called polymer brushes, a term describing strictly linear polymers that are densely grafted via one end to an interface. Such coatings display unique physical properties useful for a variety of applications. In particular, the high structural control of polymer brushes that can be realized by controlled or living polymerization techniques draws much attention. The contribution by Takeshi Fukuda et al. on high-density polymer brushes outlines the synthetic possibilities as well as the unique properties of polymer brushes. Such coatings will surely play an important role in innovative surface science and nanotechnology. The present contributions also reflect an ongoing trend: the development of defined heterogeneities on nearly any length scale. This can be realized by structured polymer coatings, gradients and control of the topography via the SIP reaction conditions. Jan Genzer's contribution on the preparation of polymer brush gradients is a good example. As it relates to defined structural variation and control of the macromolecular design of grafting polymers via SIP, I would like to point the reader to the contributions by Takehisa Matsuda on surface graft microachitectures or by David Bergbreiter discussing the synthesis and applications of hyperbranched polymers on surfaces.

Originally, the reviews were to be divided into, e.g., a *Synthesis*, *Properties* and *Application* section. Fortunately, this was not possible at all. Synthesizing a polymer coating by SIP is performing materials science from scratch. Introducing a slightly different monomer or changing the solvent will automatically alter the properties of the surface such as its wetting behavior, topography, elasticity, homogeneity, etc. It is exciting (and difficult!) to characterize the layers and find out why an altered reaction condition had such an impact upon the various layer properties. Thus, the researcher is immediately involved in various aspects of surface science and analytical challenges. This is reflected in all contributions. For example, Daniel Dyer discusses the fundamental and interesting aspect of the photoinitiated synthesis of polymer brushes. Of course, the enormous advances in surface-sensitive characterization techniques developed for the investigation of self-assembled monolayers have provided the proper tools. However, as polymers are flexible, the investigation of the dynamic behavior of polymer coatings adds another dimension. The contribution by William Brittain on stimuli-responsive films gives an idea of the complex behavior of polymer brushes.

Besides the analytical techniques, the theoretical description of polymer brushes allows a deeper understanding of the complex dynamic behavior of polymers on surfaces and is useful for future developments. Here, Roland Netz gives – also for the non-expert – a very helpful theoretical background on the theoretical approaches for the description of neutral and charged polymer brushes.

The interest in polymer brushes and defined polymer coatings prepared via SIP is not at all restricted to the polymer community or the surface science community. The demand for tailored, functionalized and adaptive surfaces comes from a multitude of scientific branches and also from industry. Possible applications are already discussed in many of the contributions compiled here. Besides polymer science, surface chemistry and physics, they include catalysis, biomedical applications, microfluidics and nanotechnology. This creates a highly interdisciplinary, lively and fruitful environment.

Finally, I would like to thank all authors for their time and effort to make a state-of-the-art overview of surface-initiated polymerization possible. An edited book is only as good as its contributions and I had the privilege to compile contributions of the highest quality.

I am also grateful to Ms. Ulrike Kreusel and Dr. Marion Hertel from Springer for their professional help and patience.

Munich, January 2006 *Rainer Jordan*

Contents

Structure and Properties of High-Density Polymer Brushes Prepared by Surface-Initiated Living Radical Polymerization
Y. Tsujii · K. Ohno · S. Yamamoto · A. Goto · T. Fukuda 1

Photoinitiated Synthesis of Grafted Polymers
D. J. Dyer 47

Photoiniferter-Driven Precision Surface Graft Microarchitectures for Biomedical Applications
T. Matsuda 67

Polymer Brushes by Anionic and Cationic Surface-Initiated Polymerization (SIP)
R. Advincula 107

Metathesis Polymerization to and from Surfaces
M. R. Buchmeiser 137

Author Index Volumes 101–197 173

Subject Index 199

Contents of Volume 198

Surface-Initiated Polymerization II

Volume Editor: Rainer Jordan
ISBN: 3-540-30251-4

Hyperbranched Surface Graft Polymerizations
D. E. Bergbreiter · A. M. Kippenberger

**Surface-Grafted Polymer Gradients:
Formation, Characterization, and Applications**
R. R. Bhat · M. R. Tomlinson · T. Wu · J. Genzer

**Surface Rearrangement of Diblock Copolymer Brushes
– Stimuli Responsive Films**
W. J. Brittain · S. G. Boyes · A. M. Granville · M. Baum · B. K. Mirous ·
B. Akgun · B. Zhao · C. Blickle · M. D. Foster

Theoretical Approaches to Neutral and Charged Polymer Brushes
A. Naji · C. Seidel · R. R. Netz

Structure and Properties of High-Density Polymer Brushes Prepared by Surface-Initiated Living Radical Polymerization

Yoshinobu Tsujii · Kohji Ohno · Shinpei Yamamoto · Atsushi Goto · Takeshi Fukuda (✉)

Institute for Chemical Research, Kyoto University, Uji, 611-0011 Kyoto, Japan
tsujii@scl.kyoto-u.ac.jp, ohno@scl.kyoto-u.ac.jp, shinpei@msk.kuicr.kyoto-u.ac.jp, agoto@scl.kyoto-u.ac.jp, fukuda@scl.kyoto-u.ac.jp

1	Introduction	3
2	Surface-Initiated Living Radical Polymerization	5
2.1	Surface-Initiated ATRP	8
2.2	Surface-Initiated NMP	15
2.3	Surface-Initiated RAFT Polymerization	15
3	Structure and Properties of High-Density Brushes	17
3.1	Swollen Brushes	17
3.1.1	Conformation of Graft Chains	17
3.1.2	Properties of Graft Chains	22
3.2	Dry Brushes	24
3.2.1	Glass Transition	25
3.2.2	Mechanical Properties	27
3.2.3	Miscibility with Polymer Matrix	28
4	Precise Design for Functional Surfaces	29
4.1	Controlled Grafting of Functional Polymers	29
4.2	Morphological Control	31
5	Applications of High-Density Polymer Brushes	33
5.1	Fine Particles Coated with High-Density Polymer Brushes	33
5.2	Application as Novel Biointerface	37
6	Conclusion and Prospect	39
References		40

Abstract Surface modifications by polymers are becoming increasingly important for various applications ranging from biotechnology to advanced microelectronics. Recent successful applications of living radical polymerization (LRP) made it possible to graft various low-polydispersity polymers including simple homopolymers, end-functionalized polymers, block/random/gradient copolymers, and functional polymers. At the same time, this technique has brought about a striking increase of graft density. Graft chains in such a high-density polymer brush were found to be highly extended in good solvent, even to the order of their full lengths. It was also found that a high-density polymer brush has characteristic properties, in both swollen and dry states, quite different and unpredictable from those of the semi-dilute or moderately dense polymer brushes previously studied. This review highlights the recent development of surface-initiated LRP and the structures, properties, and potential applications of thereby obtainable high-density polymer brushes. It is believed that surface-initiated LRP is opening up a new route to "precision" surface modification.

Keywords Graft polymerization · Living radical polymerization · Polymer brush · Surface modification · Tethering polymer

Abbreviations

Φ	volume fraction of polymer segment
AAm	acrylamide
AFM	atomic force microscopy
AT	atom transfer
ATRP	atom transfer radical polymerization
AuNP	gold nanoparticle
CTS	2-(4-chlorosulfonylphenyl)ethyltriethoxysilane
D	separation
DC	dissociation-combination
DEPN	N-$tert$-butyl-N-(1-diethylphosphono-2,2-dimethylpropyl)-N-oxy
DT	degenerative chain transfer
F	interaction force
FRP	free radical polymerization
GMA	glycidyl methacrylate
GPC	gel permeation chromatography
G_f	free energy of interactions
LRP	living radical polymerization
L_c	contour length
L_{cw}	weight-average contour length
L_d	dry thickness
L_e	equilibrium thickness
M	monomer
MMA	methyl methacrylate
MP	magnetic nanoparticle
M_n	number-average molecular weight
M_w	weight-average molecular weight
N	degree of polymerization
NMP	nitroxide-mediated polymerization
OEGMA	oligo(ethylene glycol) methyl methacrylate
P·	polymer radical
P-X	dormant species
P4VP	poly(4-vinylpyridine)
PAAm	poly(acrylamide)
PDMA	poly(N,N-dimethylacrylamide)
PEMA	1-propoxyethyl methacrylate
PEO	poly(ethylene glycol)
PMAA	poly(methacrylic acid)
PMMA	poly(methyl methacrylate)
PNIPAM	poly(N-isopropylacrylamide)

PS	poly(styrene)
R	radius of the probe sphere
RAFT	reversible addition-fragmentation chain transfer
SAM	self-assembled monolayer
ST	styrene
SiP	silica particle
Sp	spartein
TEMPO	2,2,6,6-tetramethyl-1-piperidinyloxy
TERP	organotellurium-mediated radical polymerization
T_g	glass transition temperature
X	capping agent
a^2	cross-sectional area per monomer unit
dHbipy	4,4′-diheptyl-2,2′-dipyridyl
dNbipy	4,4′-dinoyl-2,2′-dipyridyl
l_0	chain length per monomer unit
pK_a	negative logarithm of acid dissociation constant
v_0	molecular volume per monomer unit
z	distance from the substrate surface
κ_p	plate compressibility
σ	raft density
σ^*	dimensionless graft density
σ_d	surface density of dormant species
σ_i	initiator density

1
Introduction

Polymers end-grafted on a solid surface play an important role in many areas of science and technology, e.g., colloidal stabilization, adhesion, lubrication, tribology and rheology [1–5]. The conformation of those polymers in a solvent can dramatically change with graft density; [6–8] at low graft densities, they will assume a "mushroom" conformation with the coil dimension similar to that of ungrafted (free) chains. With increasing graft density, graft chains will be obliged to stretch away from the surface, forming the so-called "polymer brush". Polymer brushes may be categorized into two groups different in graft density. One is the "semi-dilute" brush, in which polymer chains overlap with each other but their volume fraction is still so low that the free energy of interaction can be approximated by a binary interaction, and the elastic free energy, by that of a Gaussian chain. Structure and properties of semi-dilute polymer brushes are relatively well understood both experimentally and theoretically. Theoretical analyses [9, 10] predicted that the equilibrium thickness (L_e) of the semi-dilute polymer brush in good solvent varies like

$$L_e \propto N\sigma^{1/3}, \tag{1}$$

where N and σ are the degree of polymerization and the surface density of the graft chains, respectively. The most important feature of this expression is that L_e depends on N in a linear way, while the dimension of an isolated chain in good solvent is scaled as $N^{3/5}$. This means that the graft chains adopt a stretched conformation. Extensive efforts have been made to characterize the semi-dilute polymer brushes systematically prepared by the "grafting-to" method using end-functionalized polymers or block copolymers with the terminal group or one of the blocks selectively adsorbed on the surface. For example, the brush height and the segment density profile in good solvent were studied by neutron reflectometry [11–15], and the interaction forces between polymer brush surfaces were directly measured by a surface force apparatus [9, 16–18] and an atomic force microscope (AFM) [10, 19, 20]. These studies have supported the theoretical predictions mentioned above.

The other category is the "concentrated" brush or the high-density brush, for which the above-mentioned approximations are no longer valid and higher-order interactions should be taken into account. In this regime, therefore, graft chains are expected to exhibit different properties from those in the semi-dilute regime due to higher-order interactions between graft chains. Theoretical analyses taking account of these interactions predicted that the repulsive force will much more steeply increase with increasing graft density [21, 22]. However, the structure and properties of high-density polymer brushes could not be well studied experimentally because of the difficulty of preparing well-defined high-density brushes. The above-mentioned grafting-to method gives limited graft densities due to the concentration gradient built up by the already grafted chains [23]. Namely, once the surface is significantly covered with polymers, they give a strong kinetic hindrance against the grafting of new chains. An alternative method to prepare polymer brushes is the "grafting-from" method, that is, the graft polymerization starting with initiating sites fixed on the surface. In this technique, the addition of monomer to growing chain ends or to primary radicals is not strongly hindered by the already grafted chains in a good solvent condition. Therefore, this technique is more promising to produce a polymer film with a larger thickness and a higher graft density than the grafting-to technique [24–29]. Earlier strategies were mostly based on conventional free radical polymerization (FRP), which successfully gave grafted films of increased density. However, no clear experimental evidence has been reported to confirm the achievement of structure and properties specific to high-density brushes, which suggests that the achieved graft density may still be in the semi-dilute regime. The generally poor control of chain length and chain-length distribution by conventional FRP poses complexities in defining the prepared brushes. In a polydisperse brush, longer chains will form the outer fringe of the swollen brush with a relatively low segmental density, which may mask the possibly unique properties of the higher density portion of the brush near the substrate surface.

Recently, living polymerization techniques including anionic, cationic, ring-opening, ring-opening metathesis, and living radical polymerizations were successfully applied to surface-initiated graft polymerization to prepare

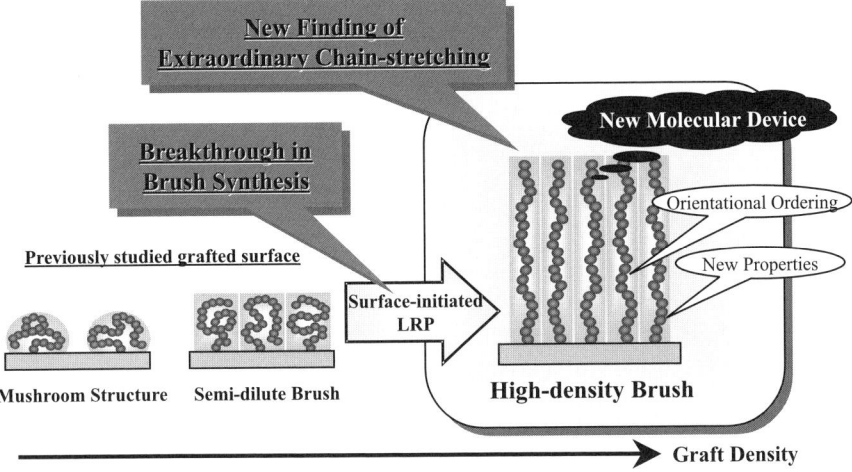

Fig. 1 Development of high-density polymer brush

a polymer brush with the basic brush parameters controlled [30]. Of these techniques, living radical polymerization (LRP) has been most widely used for its tolerance to impurities and versatility to various monomers. Surface-initiated LRP has proved to provide a dramatic increase of graft density, giving well-defined high-density polymer brushes (see Fig. 1). Recent studies have revealed that such brushes have characteristic structure and properties quite different and unpredictable from those of the semi-dilute polymer brushes previously studied.

This review article focuses on high-density polymer brushes formed by surface-initiated LRP on the surfaces of various wafers, particles, and porous materials. The next section (Sect. 2) will deal with controlled synthesis of high-density polymer brushes. Section 3 will be concerned with experimental studies characterizing their structure and properties in solvent-swollen and dry states. Section 4 will discuss the precise design for functional surfaces. Finally, potential applicabilities of high-density polymer brushes will be discussed in Sect. 5. Another family of high-density polymer brushes, i.e., "molecular brushes" including comb-like graft and star polymers have also been successfully synthesized by use of LRP, for which the reader is referred to recent reviews [31–34].

2
Surface-Initiated Living Radical Polymerization

LRP has attracted enormous attention over the past decade for providing simple and robust synthetic routes to well-defined, low-polydispersity polymers [33–56]. The basic concept of LRP is a reversible activation process

(Scheme 1). The dormant (end-capped) chain P-X is supposed to be activated to the polymer radical P· by thermal, photochemical, and/or chemical stimuli. In the presence of monomer M, P· will undergo propagation until it is deactivated back to P-X. If a living chain experiences the activation-deactivation cycles frequently enough over a period of polymerization time, all living chains will have a nearly equal chance to grow, yielding a low-polydispersity product. (The "living" chain denotes the sum of the dormant and active chains.) Miscellaneous capping agents X are used for LRP. Examples are stable nitroxides (nitroxide-mediated polymerization, NMP) [57–60], iodine (iodide-mediated polymerization) [61–63], halogens with transition metal catalysts (often termed atom transfer radical polymerization (ATRP)) [64, 65], dithioester and other unsaturated compounds (iniferter polymerization [66] or reversible addition-fragmentation chain transfer (RAFT) polymerization [67, 68]), tellanyls (organotellurium-mediated radical polymerization (TERP)) [69], and cobalt complexes [70]. The reversible activation reactions in most successful LRPs are classified into three types, which are (a) the dissociation-combination (DC) (for NMP and TERP), (b) the atom transfer (AT) (for ATRP), and (c) the degenerative chain transfer (DT) (for iodide, RAFT, and TERP) mechanisms.

Of these LRPs, NMP, ATRP, and RAFT polymerization have already been applied to surface-initiated graft polymerization by immobilizing either a dormant species or a conventional radical initiator on the surface. In the latter case, a capping agent is added in the solution phase (reverse LRP). Complementary review articles concerning surface-initiated LRP are now available [42, 44, 50]. The use of a surface-bound dormant species is more promising to obtain well-defined high-density polymer brushes. Figure 2 shows examples of surface-immobilizable dormant species including alkoxyamines for NMP and chlorosulfonyl-, chlorobenzyl- and haloester-compounds for ATRP. The RAFT-mediated graft polymerization was achieved mainly by the combined use of a surface-bound azo initiator and a free RAFT agent.

Surface-initiated (or surface-confined) polymerization brings about different situations from those in solution polymerization, which come from the immobilization of initiating (dormant) species and tethering and crowding of polymer chains on the surface. The following question arises: does the initiation efficiently occur on the surface? The initiation efficiency will be directly reflected on the graft density, which is one of the most important parameters of brush surfaces. Among other questions is whether the polymerization is controllable on the surface (within a graft layer) like in solution. The limited

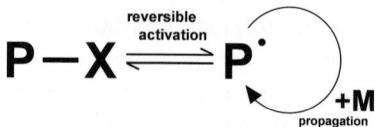

Scheme 1

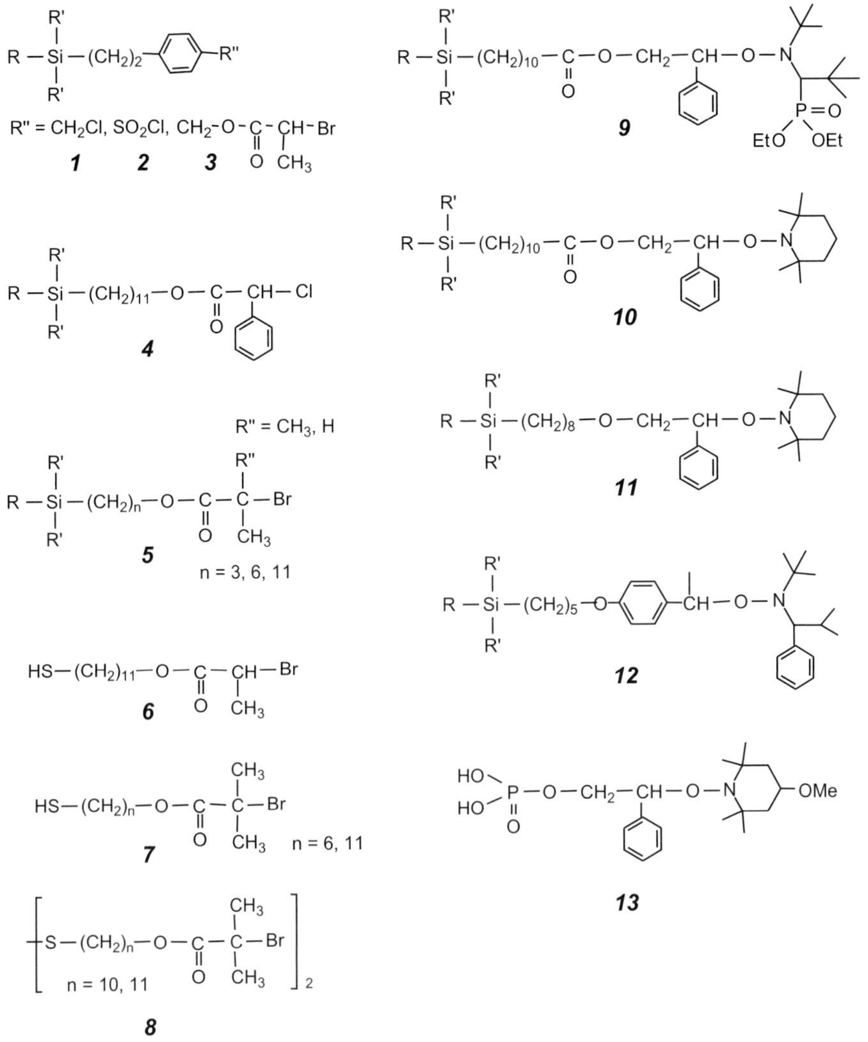

Structure and Properties of High-Density Polymer Brushes

Fig. 2 Examples of surface-immobilizable dormant species

surface area of, e.g., a flat substrate will lead to a low overall concentration of initiating (dormant) and capping species, resulting in poor control of polymerization, which must be circumvented for obtaining well-defined polymer brushes. Tethering and crowding of polymers will affect the local concentration of reactants, and hence their reaction rates. This will have the most important effect on intermolecular reactions among graft chains, e.g., the termination between graft radicals and the degenerative chain transfer between a graft radical and a graft dormant. For discussing this, it will be helpful to compare the chain length and chain-length distribution of graft polymers with those of free polymers polymerized under similar conditions; an interesting system is the polymerization with an added free initiator, since the polymerization simultaneously proceeds in both the graft-layer and solution phases. These issues, specific to surface-initiated polymerization, will be discussed below.

2.1
Surface-Initiated ATRP

ATRP has been widely applied to surface-initiated graft polymerization on a variety of materials including flat substrates [71–84], fine particles [85–108], and porous materials [109–113]. Ejaz et al. first succeeded in synthesizing a dense brush of low-polydispersity poly(methyl methacrylate) (PMMA) by the surface-initiated ATRP with copper/ligand complexes. They deposited a commercially available silane-coupling agent, 2-(4-chlorosulfonylphenyl)ethyltriethoxysilane (CTS, *2* in Fig. 2), on a silicon wafer by the Langmuir–Blodgett technique to form a covalent bond by the coupling reaction with the silanol group on the silicon surface [71]. Figure 3 schematically illustrates the graft polymerization process: the activator (A) such as a Cu^I complex abstracts the halogen atom of the immobilized initiating dormant species, e.g., CTS, or the grown dormant chain, giving a propagating radical, to which some monomer units are added until it is recapped to be a dormant chain. This cycle occurs repeatedly and randomly on the halogenated sites on the surface, thus allowing all graft chains to grow slowly and nearly simultaneously, when viewed in a longer time scale, hence in a controlled fashion. This contrasts to the conventional radical graft polymerization, in which a radical formed on the surface instantly grows to

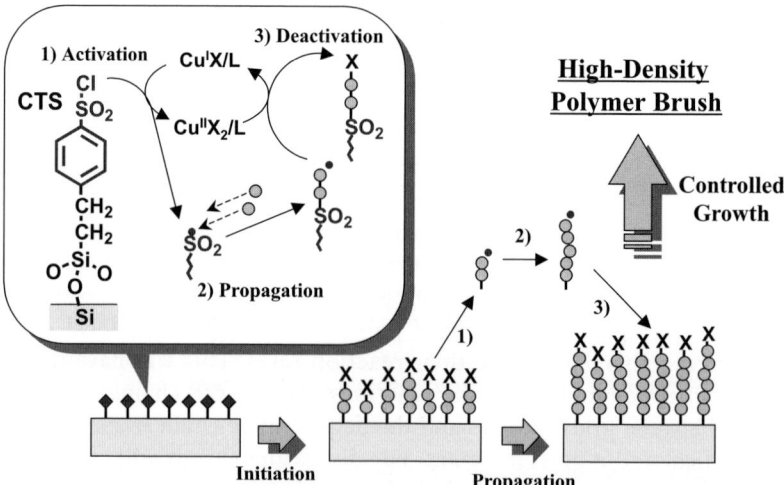

Fig. 3 Schematic illustration of surface-initiated atom transfer radical polymerization

a high-molecular-weight polymer and the graft polymerization proceeds by an increase of the number of graft chains. Hawker et al. synthesized a haloester type of silane coupling agent and successfully grafted PMMA using a Ni complex on the initiator-immobilized silicon wafer [74]. As mentioned above, the graft polymerization may be controlled only poorly because of a low overall concentration of dormant species immobilized on a flat substrate. In these cases, therefore, the free initiator was added to produce free polymers, thereby to control the graft polymerization automatically, as in the case of free (solution) polymerization controlled by the so-called persistent radical effect [51]. Another advantage of adding the free initiator is that the produced free polymer can be a useful measure of the chain length and the chain-length distribution of the graft polymer. Good agreement in the number-average molecular weight (M_n) and the polydispersity index (M_w/M_n) between the graft and free polymers has been already observed by several research groups by directly characterizing the graft polymers cleaved off the silica particles (SiPs) [74, 85, 88, 114]. Figure 4 confirms this for SiPs with a wider range of diameters (ranging from 12 nm to 1.5 μm), suggesting that the surface curvature has little effect on the growth of graft chains. In the surface-initiated ATRP with the additional free initiator, the amount (mass per unit area) of graft polymer increased proportionally to the M_n of the free polymer produced from the free initiator (see Fig. 5), meaning that the graft density was kept constant during the course of polymerization.

An alternative method to control the surface-initiated LRP with a low overall concentration of the dormant species is to add an appropriate amount of the capping agent X like $CuBr_2$ prior to polymerization. In this case, no free polymer is produced, and hence no additional process to remove the free polymer possibly in the graft layer is required. This was firstly demonstrated by ATRP on a silicon wafer by Matjyaszewski et al. [73] and subsequently

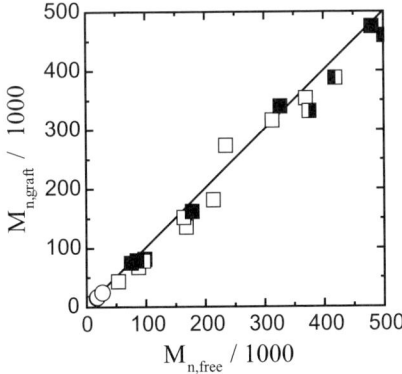

Fig. 4 Relationship between M_n of cleaved polymers and that of free polymers for silica particles with diameters of 12 nm (*circles*), 130 nm (*squares*), 290 nm (*half squares*), 790 nm (*half black squares*), and 1550 nm (*black squares*)

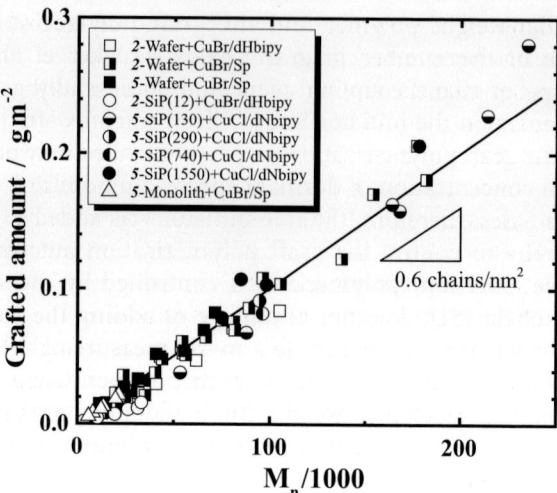

Fig. 5 Relationship between the amount of graft polymers and M_n of free polymers. The graft polymerization was carried out under various conditions on silicon wafer (*squares*), silica particles with varying diameter (d = 12, 130, 290, 740, 1550 nm) (*circles*), and silica monolith with 50-nm mesopores (*triangles*). Two types of immobilized initiators, **2** and **5** (n = 6 and R″ = CH$_3$) in Fig. 2, two types of copper halides, CuBr and CuCl, and two types of ligands, spartein (Sp) and dipyridyl derivatives (4,4′-diheptyl-2,2′-dipyridyl (dHbipy) and 4,4′-dinonyl-2,2′-dipyridyl (dNbipy)), were used

by some other researchers [75, 76, 81]. They observed a linear increase of the amount of graft polymer with polymerization time, which suggested that the graft polymerization proceeded in a living fashion with a constant graft density. This was reasonable, since the overall concentrations of the monomer and catalytic species little changed with time because of the so small specific area of a flat substrate.

A Au-coated substrate is another model surface, to which many surface characterization methods can be applied. To achieve surface-initiated ATRP on Au-coated substrates, some haloester compounds with thiol or disulfide group were developed [80–84]. Self-assembled monolayers (SAM) of these compounds were successfully prepared on a Au-coated substrate and used for ATRP graft polymerization. Because of the limited thermal stability of the S – Au bond, the ATRP was carried out at a relatively low temperature, mostly at room temperature, by using a highly active catalyst system and water as a (co)solvent (water-accelerated ATRP).

Surface-initiated ATRP was applied not only on planer substrates but also on various kinds of fine particles. The latter systems will be reviewed separately in Sect. 5.1. Porous materials are also fascinating targets for chromatographic application making use of the unique structure and properties of high-density polymer brushes. Wirth et al. were the first to report the grafting of poly(acrylamide) (PAAm) on a porous silica gel [109, 110].

Here, we focus on the surface-initiated ATRP of methyl methacrylate (MMA), which has been extensively and precisely studied. Figure 5 shows the relationship between the amount of graft polymer and the M_n of the free polymer (approximately equal to the M_n of the graft polymer; see above). These data were obtained by the authors' group under various polymerization conditions on a variety of silicate materials including silicon wafer, SiPs of differing diameters, and monolithic silica gel with ca. 50-nm mesopores [71, 89, 115–117]. These have flat, convex, and concave/convex surfaces, respectively. Detailed experimental conditions are given in the captions. In all cases, the free dormant initiator was added to control the polymerization. This figure suggests that all the data fall on the same straight line. This proportional relationship means that the graft density was constant throughout the course of polymerization, and that the graft polymerization proceeded in a living fashion. From the slope of the line, the graft density (σ) was estimated to be about 0.6 chains nm^{-2}. Table 1 summarizes the M_n, M_w/M_n, and σ values available in the published papers. The graft density is reasonably close to the value obtained in Fig. 5, ranging from about 0.5 to about 0.8 chains nm^{-2}. These values are much higher than those obtained by the conventional techniques. It can be concluded that surface-initiated ATRP afforded not only the control of chain length and chain-length distribution of graft chains but also higher graft densities. The question arises as to why such a high graft density was obtained. The most important difference between LRP and conventional FRP might be the initiation efficiency. It should be essentially 100% in LRP. This must be so, since an LRP run usually experiences tens to hundreds of activation-deactivation cycles to give a low-polydispersity polymer, and the initiation reaction in LRP is just one of these activation processes in essence [39, 118].

Nevertheless, these graft densities are smaller than the density of the initiating dormant species (σ_i) on the surface in many cases, suggesting lower initiation efficiency of surface-initiated ATRP than that of the solution system. As a possible reason for this, one may argue that at an early stage of polymerization, graft radicals will be strictly localized near the substrate surface and effectively terminated with each other, and only the surviving chains will grow up to longer chains. However, this is unlikely, because the graft density of PMMA brushes seems to be independent of polymerization conditions and hence the concentration of active radicals. Figure 5 and Table 1 contain the data obtained for different types of dormant species, copper halides, and ligands and different types of surfaces (flat, convex, and concave/convex) and different kinds of materials (silicate and gold). Interestingly, Baum et al. achieved nearly the same graft density by the RAFT system. Another possible reason to explain the relatively low initiation efficiency of surface-initiated ATRP is concerned with the excluded volume effect of the monomer on the surface (see Fig. 6). To discuss this, let us consider the dimensionless graft density σ^* defined as $\sigma^* = a^2\sigma$, where a^2 is

Table 1 Surface-initiated ATRP and RAFT polymerization of MMA

Method	Immobilized dormant	Activator (or RAFT agent)	Free initiator	Substrate	M_n	M_w/M_n	σ/chains nm^{-2}	Refs.
ATRP	5 ($n = 6$, $R'' = CH_3$)	NiBr$_2$/PPh$_3$	EBiB	Si wafer	$10000 \sim 50000$*		0.7	[74]
ATRP	2	CuBr/dHbipy	TsCl	Si wafer	$45400 \sim 83300$*	$1.1 \sim 1.3$*	$0.8 \sim 0.6$	[148]
ATRP	5 ($n = 3$, $R'' = CH_3$)	CuBr/PMDETA	EBiB	Si wafer	$30000 \sim 90000$	$1.15 \sim 1.27$*	0.3	[77]
ATRP	5 ($n = 11$, $R'' = CH_3$)	CuBr/HMTETA	EBiB	H-Si wafer	$10000 \sim 20000$*	ca. 1.2*	0.3	[79]
ATRP	6	CuBr/Me$_6$TREN	—	Au-coated wafer	$33100 \sim 68900$	$1.29 \sim 1.45$	$0.7 \sim 0.4$	[80]
ATRP	7 ($n = 11$)	CuBr/bipy	—	Au substrate	35000	1.59	0.6	[169]
ATRP	8 ($n = 10$)	Fe(II)Br$_2$/PPh$_3$	EBiB	Au-coated glass slide	$6000 \sim 60000$*	$1.1 \sim 1.4$*	0.5	[173]
RAFT	azo-initiator	2-phenylprop-2-yl (dithiobenzoate)	AIBN	Si wafer	21300* 25400*	1.1* 1.72*	0.8 0.5	[127]

* M_n and M_w/M_n values measured for free polymers

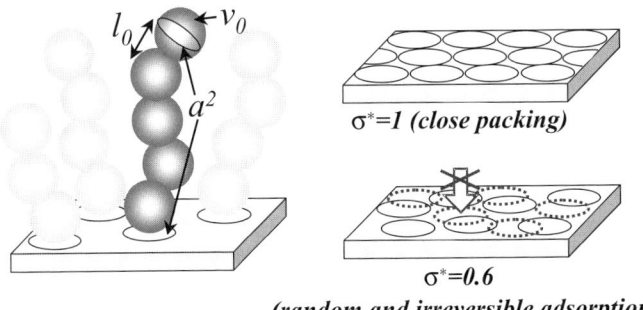

Fig. 6 Excluded volume effect of monomer on initiation efficiency (see text for detail)

the cross-sectional area per monomer unit given by $a^2 = v_0/l_0$ with v_0 being the molecular volume per monomer unit (estimated from the bulk density of monomer in this case) and l_0 being the chain contour length per monomer unit ($l_0 = 0.25$ nm for vinyl polymers). The maximum value of σ^* is unity, corresponding to the close packing of graft chains with all-*trans* conformation. The graft densities of 0.5 to 0.8 chains nm^{-2} correspond to $\sigma^* = 0.4$ to 0.6. These values are close to the average surface coverage ($\sigma^* = 0.6$) of spheres that are simulated to *randomly and irreversibly* adsorb on a flat surface in a monolayer (Tsujii et al., 2005, personal communication).

In relation to the initiation efficiency on the surface, the dependence of σ on σ_i was explored by some researchers. An almost constant graft density of PMMA brushes was observed at initiator densities higher than a critical value, as shown in Fig. 7, where the initiator density was changed by photodecomposition and its concentration was estimated by the FT-IR method [117, 119]. This figure suggests that the initiation efficiency is high and close to 100%, when σ_i is lower than a critical value equal to the maximum or cutoff graft density. On the other hand, Jones et al. reported that the graft density of PMMA brushes on a Au-coated mica almost linearly increased with increasing σ_i [83]. They changed σ_i by diluting the haloester thiol with inert undecane thiol. At 100% of haloester thiol immobilized, σ and σ_i were ca. 0.6 chains nm^{-2} and ca. 5 molecules nm^{-2}, respectively, indicating poorer initiation efficiency at the surface for any σ_i. Very recently, Ma et al. reported a similar result to the one given in Fig. 7, for oligo(ethylene glycol) methyl methacrylate (OEGMA) polymerized on a Au-coated substrate with a mixed SAM [120].

To discuss the kinetic aspects of surface-initiated ATRP, the system should be modeled as a confined polymerization, confined in the graft-layer phase. In the system producing free polymers, the polymerization will simultaneously proceed in the solution phase. The unbound reactants such as free polymer radicals, monomer, catalytic species, and other additives will be partitioned between these two phases. The polymerization is usually carried

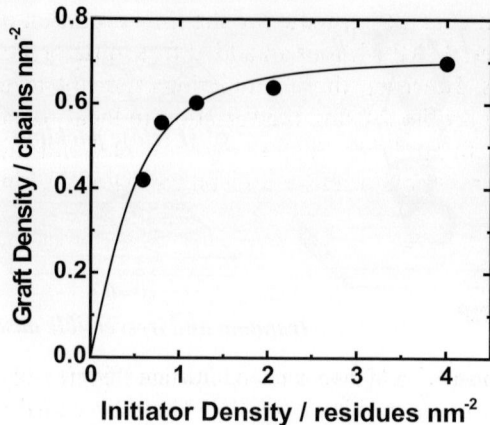

Fig. 7 Plot of graft density (σ) vs. initiator density (σ_i). The solid curve is drawn to guide the eye. Data reprocessed from [117]

out under a good solvent condition, under which the graft chains with sufficiently high σ are highly stretched, nearly to their full length, as will be discussed later. Therefore, the graft-layer phase can be assumed to have the thickness approximated by the full length of the graft chain. Then, the average concentration of polymer segments in the graft-layer phase is about σ^*, and the average concentration of graft-chain ends or graft dormants is inversely proportional to the chain length. The segmental density profile of the brush should also be taken into account. Matyjaszewski et al. [73] and Milchev et al. [121] estimated the density distributions of polymer segments and chain ends as well as the chain-length distribution by computer simulation assuming an ideal living (termination free) polymerization without activation/deactivation reactions. They suggested that the chain ends are more populated in the outer region of the brush layer, and that longer chains favorably propagate because of the gradient of polymer segment density, hence of monomer concentration within the brush, giving a broader chain-length distribution of the graft chains than the free chains produced in the solution phase. Regarding termination, one may expect an enhanced termination because of graft chain-ends being highly localized in the outer region of the brush. Experimentally, however, no definite differences in average chain length and chain-length distribution have been observed between the graft polymer and the free polymer (see above). This suggests that in both the graft-layer and solution phases, termination has a negligible or at least similar effect on the average chain length and chain-length distribution. Moreover, the termination reactions in the graft layer between graft chains as well as between graft and free chains are believed to be much reduced due to the tethering and high polymer concentration. Kinetic simulations including the activation/deactivation equilibrium and termination were presented by Xiao et al. [75] and Kim et al. [81], predicting the time evolution of grafted mount for surface-initiated ATRP without an added free initiator, comparably to experimental results. However, these simulations did not take account of the segmental density profile causing the variation in local concentrations of reactants within the brush layer. Further experimental and theoretical studies will be needed for more quantitative understanding of the kinetics of surface-initiated ATRP.

2.2
Surface-Initiated NMP

The first application of NMP to surface-initiated graft polymerization was reported by Hussemann et al. in 1999 [74]. They succeeded in densely grafting styrene (ST) using a surface-bound dormant species with a 2,2,6,6-tetramethyl-1-piperidinyloxy (TEMPO) moiety. A free dormant with TEMPO was added to control the polymerization, and the living nature of the grafting was suggested by the formation of a block copolymer brush. They observed a proportional relationship between the thickness of poly(styrene) (PS) brush on a Si wafer and the M_n of the free polymer produced from the free alkoxyamine, calculating a graft density to be ca. 0.5 chains nm^{-2}. Subsequently, Devaux et al. reported a higher graft density by a similar system excepting that the alkoxyamine moiety was immobilized by the Langmuir–Blodgett technique [122]. The controlled deposition of the initiators was claimed to play a key role to increase the graft density. In an effort to directly measure the M_n and M_w/M_n of the polymers grafted on a flat surface, they found an M_w/M_n value slightly lower, and a gel permeation chromatographic (GPC) peak molecular weight ca. 25% higher than those of the free polymers. Beyou et al. newly synthesized a silane coupling agent with a N-tert-butyl-N-(1-diethylphosphono-2,2-dimethylpropyl)-N-oxy (DEPN) moiety and applied it to grafting of PS on silica nanoparticles [123, 124]. Parvole et al. succeeded in preparing a polyacrylate brush on SiPs using a surface-bound azo-initiator along with free DEPN [125]. In an attempt to improve the dispersability of magnetite nanoparticles by grafting PS, Matsuno et al. synthesized a new alkoxyamine and successfully immobilized it on the nanoparticle surface by the reaction of a phosphonic acid group with the Fe-OH groups [126].

2.3
Surface-Initiated RAFT Polymerization

Baum et al. applied RAFT polymerization to synthesize brushes of PS, PMMA, poly(N,N-dimethylacrylamide) (PDMA), and their copolymers on azo-initiator-bound silicate surfaces [127]. 2-Phenylprop-2-yl dithiobenzoate was added as a free (unbound) RAFT agent to control the graft polymerization. Because of a very low concentration of surface-bound initiator, a free radical initiator, 2,2′-azobisisobutyronitrile, was needed to increase the polymerization rate to a practical level, like in the solution RAFT polymerization. The PMMA and PS homopolymers grafted on a SiP were cleaved and analyzed to have M_n and M_w/M_n values comparable to those of the corresponding free polymers. The graft layer grew in a linear and stepwise fashion by sequential addition of monomers to give block copolymers. Subsequently, Zhai et al. reported the synthesis of a polybetaine brush on a hydrogen-terminated silicon wafer by a similar approach using a surface-bound azo-initiator and a free RAFT agent [128].

The graft polymerization based on the RAFT process is mechanistically different from those based on other types of LRP. Since the graft chains in a high-density polymer brush are highly stretched in good solvent with their chain ends concentrated near the free surface of the graft layer (see Sect. 3), RAFT reactions would occur extraordinarily effectively among the graft polymers, and a sequence of RAFT processes may be viewed as a migration or reaction-diffusion process of the otherwise strictly localized graft radicals (Fig. 8). This surface migration of the graft radicals would certainly increase the termination reactions among them. Tsujii et al. studied the graft poly-

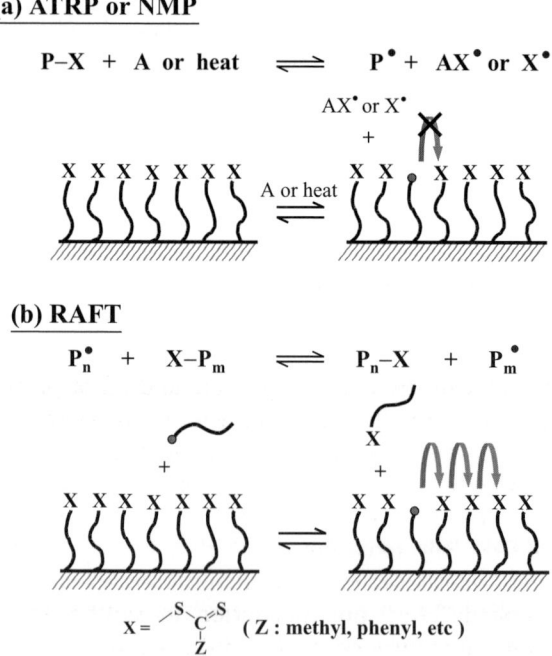

Fig. 8 Comparison of the key processes in **a** the ATRP- or NMP- and **b** RAFT-mediated graft polymerizations. Reproduced with permission from [129] (Copyright 2001 American Chemical Society)

merization of ST on the SiP grafted with probe oligomeric PS chains with a terminal dithiobenzoyl (X) group introduced by surface-initiated ATRP followed by an exchange reaction of the terminal halogen atom to the X group [129]. After the chain-extension polymerization of ST, the graft chain was cleaved from the SiP by treatment with HF and characterized. Polystyryl radicals are predominantly terminated by recombination to give dead chains of doubled molecular weight. The GPC analysis has shown two facts. First, the chain-length distribution of the main-peak component, which could be assigned to living or unterminated chains, was narrower than that of the free polymer. This suggested more frequent occurrence of the RAFT reaction on the surface than in the solution. Second, the minor-peak component, which could be assigned to the terminated chains of doubled molecular weight, was unusually large in quantity. This means a fast migration of the graft radicals on the surface. A critical value of the surface graft density below which the surface migration of the graft radicals hardly occurred was observed. Below this limit, a RAFT system would behave similarly to, for example, an ATRP system.

3
Structure and Properties of High-Density Brushes

3.1
Swollen Brushes

3.1.1
Conformation of Graft Chains

Yamamoto et al. have made an AFM study on high-density PMMA brushes prepared on a silicon wafer by surface-initiated ATRP and swollen in toluene [116, 117]. The interaction force (F) between the graft layer and a silica probe attached on the AFM cantilever was measured as a function of separation (D) between the silicon substrate and silica probe surfaces (see Fig. 9). The measured force F can be reduced to the free energy of interactions (G_f) between two parallel plates according to the Derjaguin approximation [130], $F/R = 2\pi G_f$, where R is the radius of the probe sphere (10 μm). Figure 9 shows a typical F/R vs. D curve. The true distance D between the substrate surface and the silica probe, which is usually difficult to define in AFM experiments, was successfully determined by AFM scanning the sample surface across the boundary of a scratched and an unscratched region of it [116]. A notable feature of the F/R vs. D curves is a rapid increase of the repulsive force with decreasing D. The observed repulsive forces originate from the steric interaction between the solvent-swollen brush and the probe sphere.

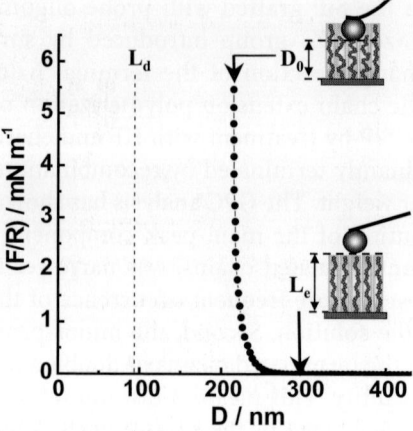

Fig. 9 Typical F/R vs. D curve between the PMMA brush (L_d = 87 nm, M_n = 121 700, M_w/M_n = 1.39) and the silica probe (attached on an AFM cantilever). The *arrowheads* indicate critical distances: L_e is the equilibrium thickness at which a repulsive force is detectable, and D_0 is the offset distance beyond which the brush was no more compressible

The equilibrium thickness (L_e) of the solvent-swollen brushes was defined as the critical distance from the substrate surface beyond which no repulsive force was detectable (cf. Fig. 9). As mentioned in Sect. 1, the scaling and self-consistent mean-field approaches predict that L_e varies like $L_e \propto L_c \sigma^{1/3}$, where L_c is the contour length of the graft chain. This relationship was confirmed by other theoretical calculations as well as by experimental data. PMMA brushes with nearly equal density and differing chain length followed this proportional relationship [116]. More interesting is the change in L_e as a function of graft density [117]. A series of PMMA brushes with the same chain length and differing graft density (0.07 < σ(chains nm^{-2}) < 0.7) were prepared by the photodecomposition of the surface initiator followed by the ATRP grafting. Figure 10 shows the plot of L_e/L_{cw} vs. σ^* in logarithmic scale, where L_{cw} is the weight-average full length of the graft chain (in all-*trans* conformation). The weight average, rather than the number average, was adopted by referring to the studies of Milner et al. [131, 132]. Figure 10 shows that L_e/L_{cw} increases with increasing σ^*, meaning that the graft chains get more and more extended as the graft density increases. For the brush with σ^* = 0.4 (σ = 0.7 chains nm^{-2}), the value of L_e/L_{cw} even approaches 0.9. This surprisingly large value certainly indicates that the graft chains in this brush are extended to an extraordinarily high extent, and compares with fully extended chains. The slope of the curve in Fig. 10 is much larger than 1/3 theoretically and experimentally expected for the semi-dilute brush (shown by the dashed line in the figure), suggesting that the scaling theory derived for the moderate-density regime is no more applicable to the present "high-density" brushes. The σ-dependency of L_e varies like $L_e/L_c \propto \sigma^n$ with n increasing

Structure and Properties of High-Density Polymer Brushes

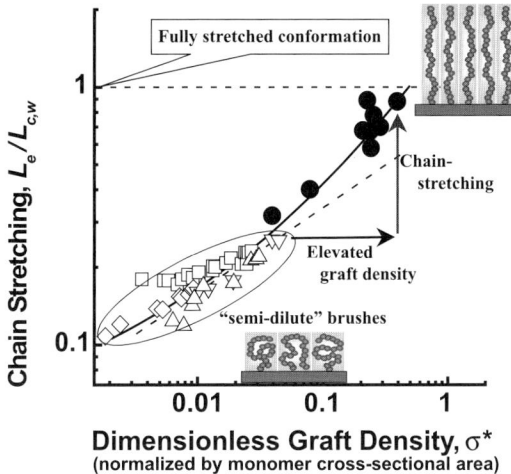

Fig. 10 Plots of L_e/L_{cw} vs. dimensionless graft density σ^*; (1) PS brushes prepared by adsorption of PS-polydimethylsiloxane block copolymers: □ ($M_{w,PS} = 60\,000$) and ◇ ($M_{w,PS} = 169\,000$) [21, 22]. (2) PEO brushes prepared by adsorption of PEO-PS block copolymers: △ ($M_{w,PEO} = 30\,800$) and ▽ ($M_{w,PEO} = 19\,600$) [201]. (3) PMMA brushes prepared by surface-initiated ATRP: ● ($M_w = 31\,300 \sim 267\,400$). Data reprocessed from [116, 117]

from about 1/3 to 1/2 with increasing σ. An increase of n in the high density regime has been predicted by the theory in which higher-order interactions were taken into account [21, 22].

A combinatorial approach was presented by Wu et al. in order to study the structure of wet polymer brushes [133, 134]. The initiating dormant species was immobilized on a silicon wafer with a gradual variation of surface density by the vapor-diffusion method. The ATRP of acrylamide (AAm) was carried out using a copper/ligand complex on the functionalized wafer, successfully giving a PAAm-grafted surface with a gradient in graft density ranging from 0.001 to 0.2 chains nm^{-2}. The M_w and M_w/M_n values of the graft polymer were estimated to be ca. 17 000 and 1.7, respectively, by cleaving and analyzing the graft polymer prepared on SiPs under the same condition. Here, the graft polymer was reasonably assumed to have the same length independent of the initiator density. The ellipsometric mapping of the brush height in dry and wet states clearly revealed the gradient brush structure on the wafer. Figure 11 shows the height in water, a good solvent, as a function of graft density proportional to the dry thickness in this case. This figure clearly indicates the mushroom-to-brush crossover at around $\sigma = 0.065$ chains nm^{-2}, beyond which the height of the wet brush was scaled as σ^n with n slightly larger than 1/3. These data cannot be directly compared with those in Fig. 10 because of the broader chain-length distribution and the lower graft-density regime of the former brushes.

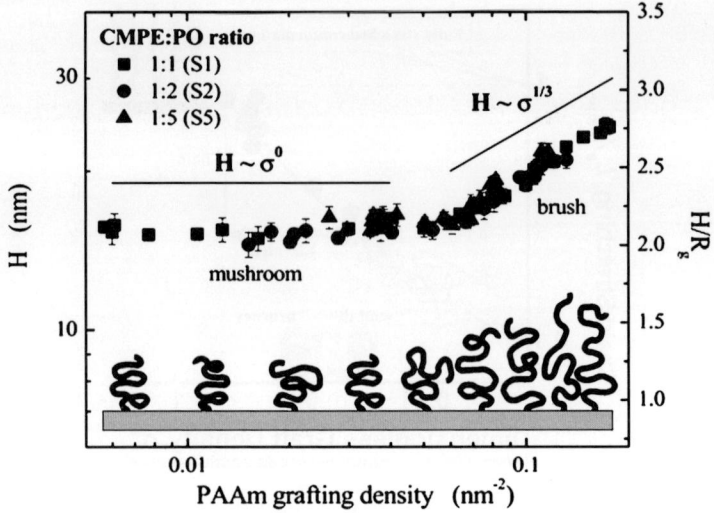

Fig. 11 Wet thickness (H) of PAAm in water as a function of the PAAm graft density for samples prepared by surface-initiated ATRP on substrates with gradient of initiator density. The initiator was immobilized by the vapor-diffusion method using mixtures of 1-trichlorosilyl-2-(m/p-chloromethyl phenyl)ethan: n-octyl trichlorosilane (w/w); 1 : 1 (*squares*), 1 : 2 (*circles*), and 1 : 5 (*triangles*). The *inset* shows a cartoon illustrating the polymer behavior. Reproduced with permission from [134] (Copyright 2003 American Chemical Society)

The above-mentioned stretching of graft chains in high-density brushes will localize graft-chain ends near the outermost (free) surface of the swollen layer. Some theories and simulations predict that this trend becomes more and more pronounced with increasing graft density [121, 135]. To verify this, a short poly(4-vinylpyridine) (P4VP) hydrophilic segment with $M_{n,P4VP} \approx 2700$ was introduced, as a probe segment, at the chain ends of a high-density PMMA brush with $M_{n,PMMA} = 22\,000$ and $\sigma = 0.62$ chains nm^{-2} by the ATRP block copolymerization, and the AFM force measurements were carried out in toluene, which is a good solvent for PMMA and a nonsolvent for P4VP (Yamamoto, Tsujii, Fukuda, 2005, personal communication). The brushes with and without terminal P4VP segments gave nearly the same advancing-mode force profiles on compression by the silica probe sphere. This suggests that the introduced P4VP segment is short enough to have little effect on the equilibrium thickness and repulsion forces against compression. More interesting were the force profiles measured in the retracting mode, in which a strong attractive force was observed at large separations for the brush with P4VP segments but not for the precursory PMMA brush (Fig. 12). Such a long-range attractive force can be attributed to the bridging of graft chains between the probe sphere and the substrate; the P4VP segment would be adsorbed on the hydrophilic surface of the silica sphere, resulting in an attractive force due to

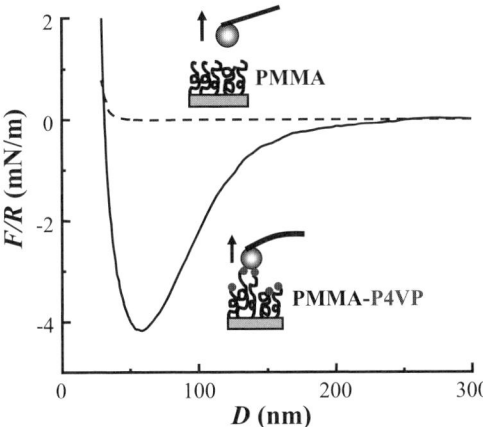

Fig. 12 Retracting-mode force profiles of high-density PMMA brushes with and without P4VP terminal segment measured in toluene using a hydrophilic silica probe sphere

an extension of graft chain in the retracting cycle, while the PMMA chains would not. On the other hand, an entirely reverse phenomenon was observed, when the silica probe was treated to give a hydrophobic surface: in this case, the PMMA brush exhibited an attractive force in the retractive mode, while the PMMA-P4VP brush gave no attractive force at all. This confirms that the P4VP segments are localized near the free surface, forming an outermost surface layer. In other words, an effective modification in chemical properties without a significant change in their physical properties can be achieved by the introduction of a short terminal block for the swollen, high-density brush surface. Such a concept specific to the high-density brush can be useful for the application of a biointerface as will be discussed later.

Other structural studies were done for the brushes with thermally responsive properties. Balamurugan et al. investigated the thermally induced hydration of poly(N-isopropylacrylamide) (PNIPAM) brush synthesized by surface-initiated ATRP on a Au substrate [136]. Surface plasmon resonance spectroscopy revealed the hydration transition occurring over a broad range of temperatures, while the contact angle measurements suggested a sharp transition. These results were interpreted by taking account of the density profile such that the polymer segment is highly solvated in the outermost region but densely packed, less solvated within the brush. Kizhakkedathu et al. synthesized the PS latex particle with brushes of PNIPAM, PDMA and poly(methoxyethylacrylamide) by surface-initiated aqueous ATRP and characterized their M_n, M_w/M_n and σ after cleaving the graft polymers from the latex surface by hydrolysis [107, 137]. The hydrodynamic thickness of the PNIPAM brush was measured by a particle-size analyzer, and it was found to be scaled as $L_c^{0.66}$ at a constant graft density. A broader range of transition temperatures was also observed. These particles were physisorbed on a glass substrate, and the interaction forces between the brush and a silicon nitride tip were measured by AFM as a function of σ and M_n. The PDMA brushes exhibited a long-range attractive force due to bridging at $\sigma = 0.012$ chains nm^{-2} and only repulsive force at higher σ with the force increasing with M_n and σ, when the brush is compressed by the AFM tip. The critical density above which no bridging force was detected was larger for the PNIPAM brush. These σ values may be for the semi-dilute brush regime.

3.1.2
Properties of Graft Chains

The breakdown of the semi-dilute brush theory was also revealed in the force-distance curve of the high-density PMMA brush. Using the scaling approach [138], de Gennes derived the equation concerning the interaction force between two parallel plates with a semi-dilute polymer brush layer, predicting that the force-distance profiles should be scaled by plotting $(F/R)\sigma^{-3/2}$ against D/L_e. In fact, the results for the previously studied semi-dilute brushes were nearly consistent to this scaling theory. High-density PMMA brushes ($\sigma > 0.4$ chains nm^{-2}) prepared by surface-initiated ATRP, however, were poorly represented by this scaling theory [116, 117]. With increasing L_c and σ, the scaled force curve increased more steeply than expected by the scaling theory, meaning that the brush layer is more resistant to compression. The strong resistance to compression is characteristic of high-density brushes, which is expected to give better dispersability of nanoparticles (see Sect. 5).

Because of the highly stretched conformation and strong resistance against compression, high-density polymer brushes are expected to give unique interactions with solute molecules in solutions. Figure 13 shows possible interactions: (a) primary adsorption at the substrate surface, (b) secondary adsorption on the brush surface, and (c) tertiary adsorption with graft chains in the brush [139]. When a solute molecule is large enough as compared with the distance between graft chains, the size exclusion may occur, preventing the primary and tertiary adsorption. This may be one of the unique features of high-density polymer brushes. In addition, theory predicts that the segmental density profile of a swollen graft film becomes steeper or sharper as the graft density increases. Therefore, such a size-exclusion limit will be set for a lower molecular weight with increasing surface density. To demonstrate this chromatographically, a high-density PMMA brush with $M_n = 15\,000$, $M_w/M_n = 1.2$, and $\sigma = 0.6$ chains nm^{-2} were prepared on the inner surface of a silica monolith column, which had a through-pore with a diameter of a few μm and a continuous silica skeleton with ca. 50-nm mesopores (He et al., 2005, personal communication). Because of the large porosity and the large surface area, such a monolith with a bimodal pore structure has been extensively investigated as a new column system of high-performance liquid

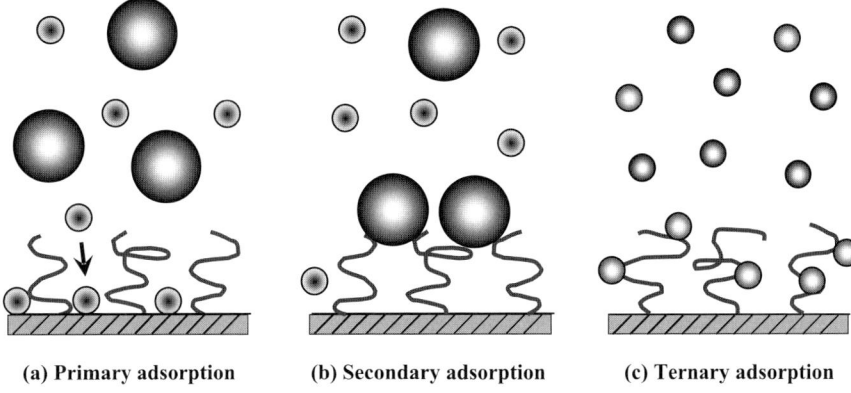

Fig. 13 Schematic illustration of possible interactions with polymer brushes

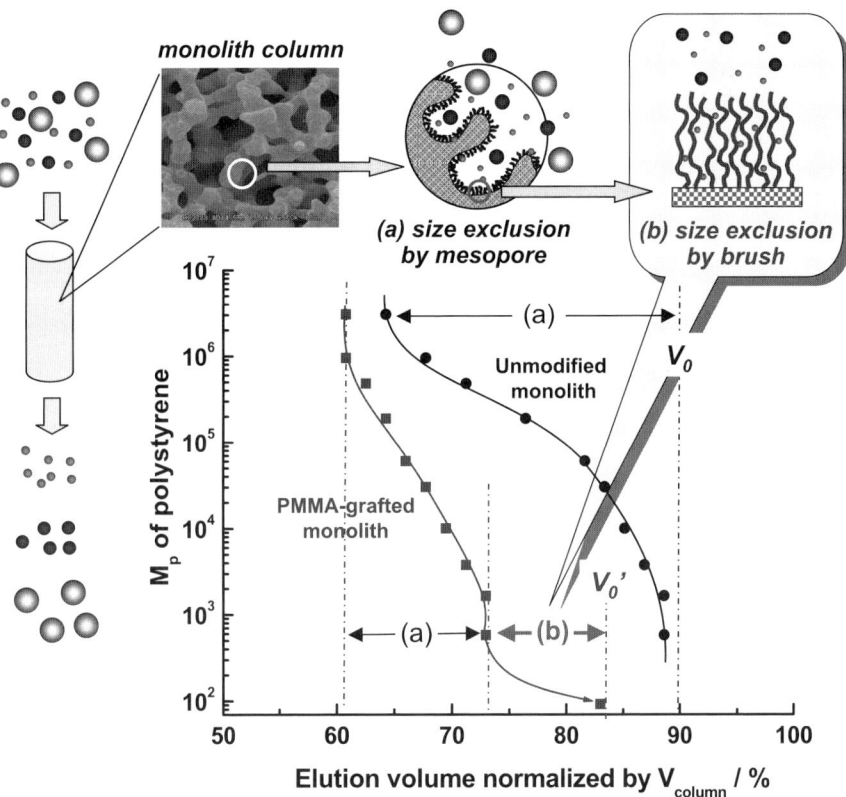

Fig. 14 Size-exclusion chromatograms of standard PSs for monolith columns with and without the high-density PMMA brush ($M_n = 15\,000$, $M_w/M_n = 1.2$, $\sigma = 0.6$ chains nm^{-2}) on an inner surface. The *inset* shows a cartoon illustrating the size exclusion mode by the brush

chromatography [140]. Figure 14 shows the size exclusion chromatograms for standard PSs with molecular weights ranging from 10^2 to 10^6 as a function of elution volume normalized by the total column volume. In the unmodified monolith, the size exclusion was observed by mesopores, giving an exclusion-limit molecular weight over 10^6. The PMMA-grafted monolith gave a size exclusion by mesopores in a lower range of elution volume. The most important feature of this chromatogram is a sharp separation newly observed in a molecular weight range between 10^2 and 10^3, which was ascribed to the size exclusion by the polymer brush. The dense and unique structure of the high-density brush would explain the sharp resolution and the exceptionally low molecular-weight limit of exclusion.

3.2
Dry Brushes

Ultra-thin polymer films on a solid substrate (supported films) are extremely interesting objects both scientifically and practically. Detailed knowledge of their structure and properties is essential for the design of advanced materials. Dry high-density polymer brushes are also interesting as a kind of supported film. For example, the above described high-density PMMA brushes with $\sigma = 0.7$ chain nm^{-2} should be highly anisotropic because their thickness in the dry state (L_d) reaches about 40% of the fully extended chain length,

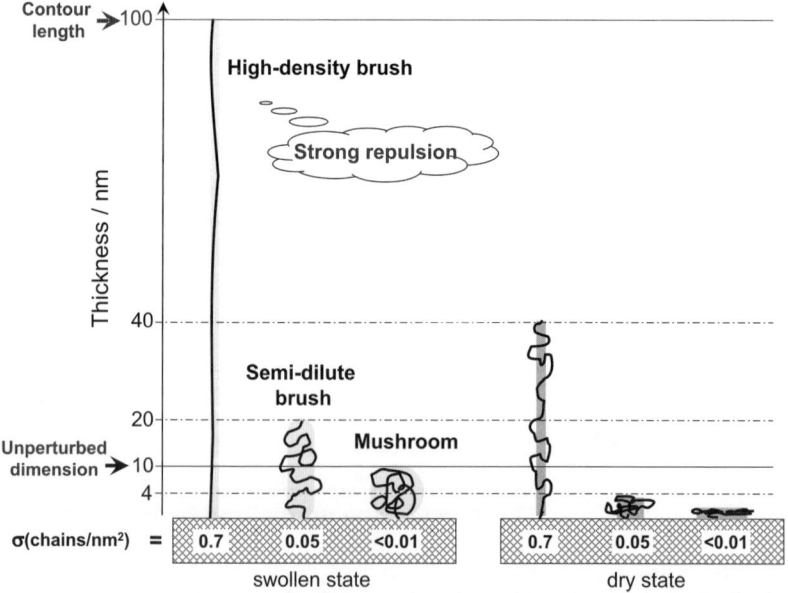

Fig. 15 Schematic illustration of conformations of end-grafted polymer chains in wet and dry states as a function of graft density

as illustrated in Fig. 15. Since the size of the unperturbed chain end-grafted on a repulsive surface is proportional to the square-root of chain length, this large value of L_d means that the chains are already highly extended in the dry state as compared with their unperturbed dimensions. This may result in structure and properties quite different from those of previously studied semi-dilute polymer brushes.

3.2.1
Glass Transition

The glass transition temperature (T_g) of a thin polymer film has been extensively studied [141–146], in which T_g was shown to strongly depend on the thickness of the film and the nature of the substrate. The first observation of the T_g of high-density PMMA brushes was made on a silicon wafer using temperature-variable spectroscopic ellipsometry [147]. Figure 16 shows the plot of the T_gs of PMMA brushes and PMMA cast films. The series of PMMA brushes studied here have a nearly constant graft density (ca. 0.7 chains nm^{-2}) and differing chain length (and hence differing dry thickness L_d). Remarkable is the difference in the T_g behavior between these two types of ultra-thin films. The molecular characteristics of the polymer forming each cast film is closely similar to those of the polymer forming the brush of (nearly) the same thickness, and therefore the T_g difference cannot be ascribed to differences in molecular characteristics such as chain length,

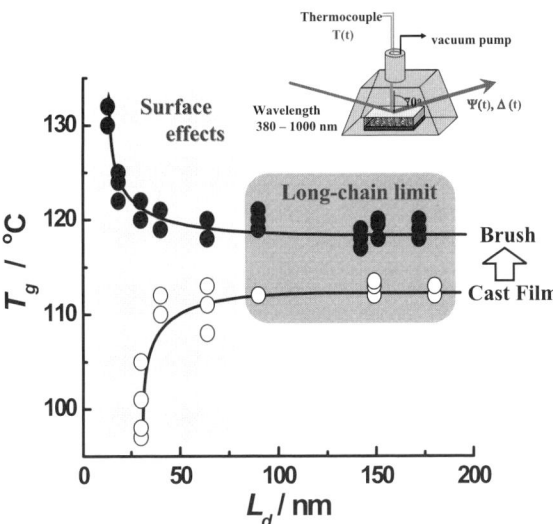

Fig. 16 L_d-dependency of T_g measured by temperature-variable ellipsometry. The *solid* and *open circles* represent the data for the brushes and the cast films, respectively. Reproduced with permission from [147] (Copyright 2002 American Chemical Society)

chain-length distribution, and stereoregularities. It may totally be ascribed to the effects of grafting, i.e., chemically binding one of the chain ends on the substrate surface.

In the range of L_d smaller than about 50 nm, cast films suffer a significant T_g depression, which was ascribed not only to the molecular weight effect but also to the interfacial effect. In contrast, the T_g of the brushes steeply increases with decreasing L_d. Obviously, end-grafting restricts the mobility of the chains. One may expect, however, that the effect of end-grafting on chain mobility would become less and less significant as the chain length increases, and in the limit of the long chain, the T_g of the graft film would become equal to that of the cast film and hence that of the bulk polymer, since all surface effects on the overall T_g of films should be unimportant in the long chain limit. Figure 16 shows, however, that this is not the case. As L_d exceeds about 50 nm, the T_gs of the brushes reach an almost constant value of about 119 °C, which is about 8 °C higher than those of the equivalent cast films. The figure strongly suggests that this difference in T_g between the brushes and cast films would be retained even in the long chain limit. This marked increase in the T_g of the long enough brushes was ascribed to an anisotropic structure/chain conformation in high-density brushes. In fact, for low- to moderate-density PMMA brushes with $\sigma \leq 0.2$ chains nm^{-2}, no T_g increment was observed [146].

Tanaka et al. have studied the surface molecular motions of PS films coated on a solid substrate by lateral force microscopy and revealed that the T_g at the surface was much lower than the corresponding bulk one [148]. Possible reasons for this included an excess free volume induced by localized chain ends, a reduced cooperativity for α-relaxation process, a reduced entanglement, and a unique chain conformation at the surface. For comparison, they examined surface relaxation behavior of high-density PMMA brushes.

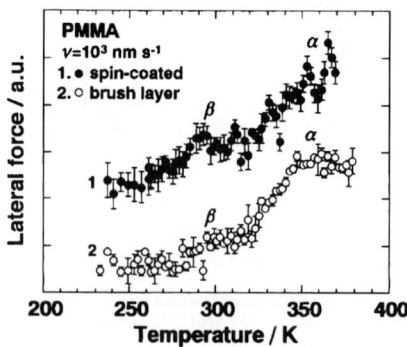

Fig. 17 Typical lateral force-temperature curves for the PMMA brush ($M_n = 45400$, $M_w/M_n < 1.2$, $\sigma = 0.8$ chains nm^{-2}) and an equivalent spin-coated film at the scanning rate of 10^{-3} nm s^{-1}. Reproduced with permission from [148] (Copyright 2003 The Society of Polymer Science, Japan)

Figure 17 shows the temperature dependence of the lateral forces measured at the scanning rate of 10^3 nm s^{-1} for a high-density PMMA brush ($\sigma = 0.8$ chains nm^{-1}) and an equivalent spin-coated film. The α-relaxation process was clearly observed accompanying a small peak, which was assigned to a surface β-process. They commented that the surface molecular motion of the brush layer possibly differs from that of the spin-coated film but that it was rather difficult to conclude this because of slightly scattered data.

The effect of tethering and conformational constraint on the glass transition of polymer brushes is expected to depend on the geometry of surfaces. By surface-initiated ATRP, Savin et al. prepared high-density PS brushes on a SiP with an average diameter of ca. 20 nm, and studied them by differential scanning calorimetry to find that the T_g of the brush sample with $M_n = 5230$ was 13 K higher than the ungrafted polymer with nearly the same molecular weight, but that the T_g difference reduced to ca. 2 K for the sample with $M_n = 32\,670$ [149]. These results suggest that the effect of conformational constraint was mitigated for segments residing farther away from the immobilized surface, which contrasts to the case with the flat surface mentioned above.

3.2.2
Mechanical Properties

By electromechanical interferometry, it was demonstrated that the anisotropic structure of high-density brush films is clearly reflected on their elastic properties [150]. Changes in the thickness of a graft film induced by an applied electric field (electrostriction) were measured by a Nomarski optical in-

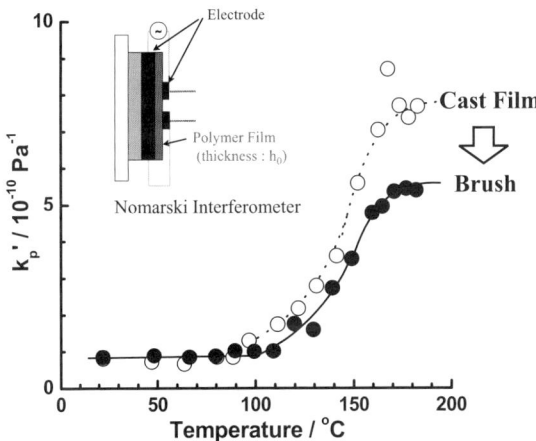

Fig. 18 Plate compressibility (κ_p) as a function of temperature (T) for the PMMA brushes ($M_n = 103\,000 \sim 214\,000$, $M_w/M_n \approx 1.2$, $\sigma \approx 0.5$ chains nm^{-1}) and the spin-coated PMMA layer ($M_n = 106\,000$, $M_w/M_n = 1.16$, $L_d = 110$ nm). The modulation frequency is 9930 Hz. Reproduced with permission from [150] (Copyright 2002 American Chemical Society)

terferometer as a function of temperature [151–153]. The analysis of the electromechanical and dielectric data yielded the plate compressibility (κ_p) of the brushes in the glassy and molten states (see Fig. 18). Comparison of the results for a high-density PMMA brush and an equivalent spin-coated PMMA layer revealed differences in elastic and dielectric properties between them. In the glassy state, there was no appreciable difference in κ_p, whereas in the molten state, κ_p of the brush was markedly (ca. 30 to 40%) lower than that of the spin-coated layer. This proved that the molten high-density PMMA brush is more resistant against compression than the equivalent PMMA melt. The low compressibility of the molten high-density brushes was attempted to be interpreted in terms of a rubber elasticity theory of a stretched polymer network with entanglements. This analysis suggested that the low compressibility is mainly attributed to a strain-hardening effect of the highly stretched entangled chains, and that there exists a considerable amount of entanglements among different graft chains contributing to the elastic modulus.

3.2.3
Miscibility with Polymer Matrix

Another example showing unique properties of high-density brushes concerns the miscibility of the polymer brush with a chemically identical polymer matrix [154]. Neutron reflectometry was applied to a series of deuterated PMMA (PMMA$_d$) brushes with a constant chain length (M_n = 46 000, M_w/M_n = 1.08) and differing graft density ($\sigma \approx 0.7$ and 0.06 chains nm^{-2}),

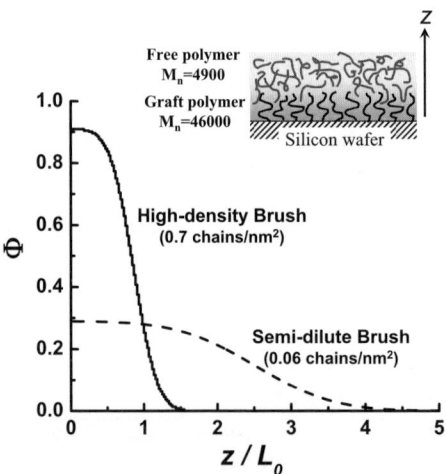

Fig. 19 Density profiles of polymer brushes faced with chemically identical polymer matrix. The polymer brushes have a constant chain length (M_n = 46 000, M_w/M_n = 1.08) and differing graft density ($\sigma \approx 0.7$ and 0.06 chains nm^{-2}), and the free polymer has M_n = 4910 and M_w/M_n = 1.1

on which hydrogenous PMMA (PMMA$_h$) with differing molecular weight (2400 < $M_{n,cast}$ < 780 000) was spin-coated. After annealing at 150 °C for 5 days in vacuum, the neutron reflectivity data were collected at room temperature and analyzed to elucidate the brush concentration profile as a function of the distance (z) from the substrate surface. A representative result is shown in Fig. 19, where the brush fraction Φ is plotted against z/L_d. The figure clearly shows that the miscibility between the brush and the free polymer strongly depends on graft density. The low-density (semi-dilute) polymer brush (M_n = 46 000, σ = 0.06 chains nm^{-2}) is swollen by an oligomeric PMMA (M_n = 4910) to a thickness about four times as thick as the dry brush. On the other hand, the high-density brush (M_n = 46 000, σ = 0.7 chains nm^{-2}) is hardly swollen by the same oligomeric PMMA, maintaining its "dry brush" structure. This phenomenon has been theoretically predicted [155] but never experimentally verified before.

4
Precise Design for Functional Surfaces

4.1
Controlled Grafting of Functional Polymers

The simplicity and versatility of LRP enable us to densely graft a variety of well-defined functional polymers, such as glycopolymers, polyelectrolytes, polymacromonomers, hyperbranched polymers, and cross-linked polymers [30, 120, 156–160]. Here, we focus on polyelectrolyte brushes, which have received much interest for their expectable properties different from those of noncharged polymer brushes. Rühe et al. reviewed recent progress in the theory, synthesis and properties of a variety of polyelectrolyte-brush systems including molecular (bottle) brushes and surface brushes on solid substrates [161]. Specifically, they discussed the swelling behavior of "strong" and "weak" polyelectrolyte brushes as a function of external conditions such as the pH and ionic strength of the surrounding medium and the addition of multivalent ions. These brushes were mostly prepared on solid surfaces by the grafting-from method via conventional FRP, and may be classified into semi-dilute brushes with respect to graft density. A weak polyelectrolyte brush gave a markedly different behavior from a strong polyelectrolyte one, because the former charge density was not constant but changed as a function of the local concentration of protons inside the brush. A most notable feature was the salt-induced swelling observed for poly(methacrylic acid) (PMAA) brushes at low salt concentrations. The pH value in the brush is governed by the requirement of charge neutrality. An added salt produces a counter cation, which promotes the dissociation of the brush, keeping charge neutrality by exchange with proton in the brush. Theory predicted that the charge neutralization would result in a difference in pH between the inside and outside of the brush. The difference will be so large for the brush with a high local concentration of ionizable groups that it will give a large shift in the critical pH value beyond which brushes are charged and swollen. This is also called the effect of "charge regulation" [162]. Rühe et al. reported that the PMAA brushes synthesized by surface-initiated polymerization via conventional FRP showed the critical pH of 4 to 5, slightly higher than the pK_a value of the carboxylic acid group.

Recently, surface-initiated LRP was applied to synthesize well-defined high-density polyelectrolyte brushes. Here, we describe the swelling behavior of a high-density weak polyelectrolyte brush to demonstrate that dense grafting can give a unique property. A PMAA brush was prepared on a silicon wafer by the surface-initiated ATRP of 1-propoxyethyl methacrylate (PEMA) followed by deprotection (cf. Fig. 20) [163]. The analysis of the simultaneously formed free polymers suggested that the graft polymerization proceeded in a controlled fashion to give a well-defined high-density polymer brush with $\sigma = 0.4$ chains nm^{-2}. The FT-IR analysis revealed that the hemiacetal ester group of poly(PEMA) grafts was quantitatively deprotected by heating at 120 °C in p-xylene containing zinc 2-ethyl hexanoate as a catalyst, giving a high-density polyelectrolyte brush. When glycidyl methacrylate (GMA) was copolymerized with PEMA, the deprotection of PEMA induced cross-linking between graft chains by the reaction of the liberated carboxylic acid with the epoxide group of GMA. The swelling behavior of the polyelectrolyte brushes with and without cross-links was studied in an aqueous solution with varying acidity. Figure 20 shows the swelling ratio of these

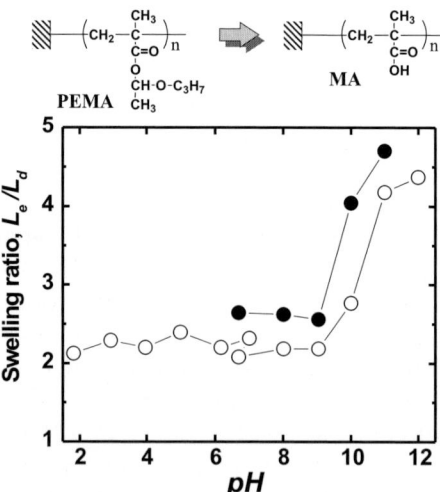

Fig. 20 Plots of swelling ratio of polyelectrolyte brushes with (•) and without (○) cross-links vs. pH of aqueous solution

brushes as a function of pH. The ratio steeply increased at around pH = 10, nearly up to the maximum possible value corresponding to the fully-stretched chain length. This means that the apparent value of the negative logarithm of acid dissociation constant (pK_a) was shifted to about 10, a value much higher than the original value ($pK_a \approx 4$) of the carboxylic acid group. It was also observed that the introduction of cross-links into the brush by the copolymerization of 3 mol % GMA enhanced the chemical stability of the graft layer even in a strong acidic/basic condition.

4.2
Morphological Control

Artificially designed fine patterning of polymer films is often demanded in various fields of science and technology such as those related to microelectronics and functional sensor devices. For this purpose, a polymer resist layer is usually spin-coated on a substrate and patterned by lithographic techniques. However, such a polymer film has a limited applicability as a functional surface because of its insufficient stability against temperature, solvents, and mechanical forces. Thus, a number of different approaches have been made to fabricate a patterned polymer layer that is stable, e.g., even in a wet system. For example, Rühe et al. prepared a patterned graft layer by selectively photoinitiating polymerization from an azo-compound chemically immobilized on a surface via conventional FRP [164]. Surface-initiated LRP allows a high-density polymer brush to grow on a patterned surface of the initiator layer by the combined use of a variety of lithographic techniques [165–176]. As well as the graft density, chain length, and chain-length distribution of the graft polymers, the morphology of the grafted surface is an important factor determining such surface properties as chemical reactivity, wettability, permeability, lubricity, biocompatibility, and electrical properties.

Surface-initiated LRP technique also makes it possible to precisely and widely control comonomer arrangement along the chain. The highly extended conformation of the graft chains in high-density brushes suggests that the grafting of block or gradient copolymers will give a layered or gradient structure stabilized by the incompatibility of different polymer segments. This will open up a new route to precise and effective tuning of surface properties. Block-copolymer brush surfaces, with a variety of characteristic surface morphologies caused by the phase separation in a nm scale, have attracted much attention for their stimulus-responsive properties [82, 177–186]. A more comprehensive review concerning this topic will be given in this thematic issue by Brittain.

Another strategy to control surface morphologies and properties by using self-assembling properties of polymer chains is to randomly graft different kinds of homopolymers on a surface. In recent years, such a "mixed homopolymer brush" has been extensively investigated both theoretically and experimentally. Some theories predicted phase diagrams consisting of various morphologies that are induced by lateral and perpendicular segregation of grafted chains in a wet or a dry state [187–191]. Those morphologies depend on structural parameters of brushes, incompatibility of the two polymers, solvent quality, and so on. Experimentally, mixed homopolymer brushes can be formed by both grafting-to and grafting-from techniques. Sidorenko et al. firstly synthesized mixed homopolymer brushes of PS and poly(2-vinylpyridine) by the grafting-from technique via conventional FRP on an azo-initiator-immobilized silicon wafer and studied their morphologies and switching properties by the treatment with selective and nonselective solvents [192]. Very recently, surface-initiated LRP was also successfully applied to precisely control the structural parameters of mixed polymer brushes. PS and PMMA were randomly grafted by the combined use of different LRPs, i.e., ATRP of MMA followed by NMP of ST [193–195]. These two LRP techniques are based on different activation mechanisms, and therefore, it was possible to independently and selectively carry out the two LRPs by controlling the temperature. Zhao immobilized a mixed SAM of two kinds of silane coupling agents with ATRP and NMP initiators and prepared mixed brushes consisting of PMMA with M_n = 26 200 and PS with varying molecular weight [193, 194]. He measured the water contact angles of these brushes after treatment with dichloromethane and observed their abrupt increase with increasing PS molecular weight. Ejaz et al. synthesized a series of PS/PMMA mixed brushes with differing composition and chain length by the similar system excepting that three-component SAM containing an inactive silane coupling agent was immobilized [195]. The addition of the inactive species enabled them to more precisely control the graft densities of the individual components in a wider range. As shown in Fig. 21, the topographic AFM studies revealed that after treatment with the selective solvent acetone, which is a good solvent for PMMA and a nonsolvent for PS, PMMA/PS-mixed homopolymer brushes with a nearly equal total graft density gave characteristic morphologies depending on the fraction of each component. With increasing the PS fraction, the surface morphology changed from a circular-domain structure to a honeycomb one via a lamellar-like one for a symmetrical brush. Depending on solvent quality, the morphology was dramatically changed: e.g., the treatment of the mixed homopolymer brush (σ_{PS} = 0.08 chains nm^{-2} and σ_{PMMA} = 0.17 chains nm^{-2}) with acetone, a selective solvent for PMMA, gave the circular-domain structure, while the treatment of cyclohexane, a selective solvent for PS, converted the surface morphology to a honeycomb structure. The height contrast was reversed depending on the selectivity of the solvent. On the other hand, no characteristic morphology was observed after the treatment with dichloromethane, a good solvent for both PMMA and PS. These morphological changes were reproducible and quite similar to those previously reported by Minko et al. [191]. Another approach was made by Zhao et al. who newly synthesized an asymmetric bifunctional silane-coupling agent

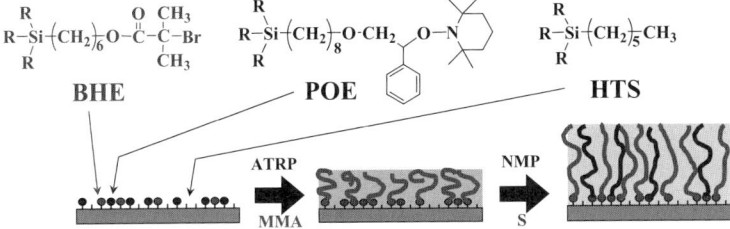

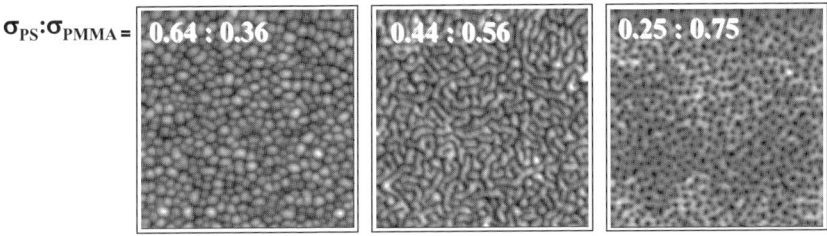

Fig. 21 Topographic AFM images of PMMA/PS-mixed homopolymer brushes with different ratios of σ_{PMMA} and σ_{PS} but nearly equal $\sigma_{PMMA} + \sigma_{PS}$ (≈ 0.5 chains nm^2) after treatment with acetone, where σ_{PMMA} and σ_{PS} are the graft densities of PMMA and PS, respectively. The *inset* schematically illustrates the sequential growth of PMMA and PS by surface-initiated ATRP and NMP on a mixed SAM of BHE, POE, and HTS

with both ATRP and NMP initiators to avoid possible preferential adsorption and cluster formation in the SAM of a multi-component system [194]. They prepared a series of PS/PMMA brushes with PMMA with a fixed M_n and PS with systematically changed M_n and observed appreciably ordered nanoscale domains after treatment with acetic acid [196].

5
Applications of High-Density Polymer Brushes

5.1
Fine Particles Coated with High-Density Polymer Brushes

This section reviews studies on the synthesis of fine particles coated with polymer brushes by surface-initiated ATRP. SiP are among the most extensively studied particles for the application of surface-initiated ATRP [85–93]. Patten et al. first succeeded in the surface-initiated ATRP of ST and MMA on two SiPs with average diameters of 75 and 300 nm [85, 86]. Several groups investigated the synthesis of hybrid SiP with different monomers. Matyjaszewski et al. synthesized an initiator-functionalized SiP with a diameter of about 20 nm via the sol-gel chemistry and grafted homo and block polymers on their surface [87, 88]. Ohno et al. also developed a new way for obtaining a monodisperse SiP coated with a high-density polymer brush with exceptionally high dispersability [89]. Most of the initiator-functionalized SiPs used in these studies are hydrophobic in character. Armes et al. reported a simple method to prepare a water-dispersible, initiator-functionalized SiP, which is based on the physical adsorption of a polyelectrolytic ATRP macroinitiator onto an anionically charged SiP [92, 93]. Ohno et al. synthesized a highly water-dispersible SiP by fixing it with an oligo(ethylene glycol) chain (Ohno et al., 2005, personal communication).

Gold nanoparticles (AuNPs) coated with a polymer brush have been synthesized by surface-initiated ATRP [94–97]. Several methods were used to introduce initiating groups onto AuNP surfaces. In the method used by Hallensleben et al., a dodecanthiol-protected AuNP was first prepared, followed by the site exchange reaction with an initiating group-carrying thiol synthesized by reduction of the corresponding disulfide [94]. In the one used by Walt et al., an undecanol monolayer-coated AuNP was first prepared, followed by the acylation of the AuNP with an initiating group-carrying acid bromide [95]. Ohno et al. prepared an initiator-coated AuNP by the simple one-pot reduction of tetrachloroaurate with sodium borohydrate or sodium citrate in the presence of an initiator group-holding disulfide [96, 97].

Magnetic nanoparticles (MPs) including Fe_3O_4, Fe_2O_3, and $MnFe_2O_4$ nanoparticles were used as a core substrate for surface-initiated ATRP [98, 99]. Yang et al. and Zhang et al. utilized the ligand-exchange reaction with an initiator group-holding acid to obtain the initiator-coated MPs [98]. Marutani et al. used an initiator-holding silane coupling agent to fix the initiator on MP, with which they prepared a polymer-coated MP with a high stability and improved dispersibility in organic solvents [99].

Patten et al. reported the synthesis of a photoluminescent hybrid particle by surface-initiated ATRP. They prepared a core-shell CdS/SiO_2 nanoparticle and modified its surface with an ATRP initiator to prepare a hybrid particle that retained the photoluminescence of the precursor CdS nanoparticle [100]. Sen et al. succeeded in the introduction of ATRP initiating sites onto an aluminum oxide particle by the reaction of an initiator-holding acid with the hydroxyl groups on the particle surface. They demonstrated that the synthesized polymer/aluminum oxide hybrid particles were useful for obtaining a polymer/ceramic composite [101]. Matyjaszewski et al. functionalized a carbon black surface with ATRP-initiating sites via the three-step reaction and succeeded in grafting poly(*n*-butyl acrylate) on the functionalized carbon black by surface-initiated ATRP [102]. Kickelbick et al. developed a novel route for the synthesis of an ATRP initiator-functionalized metal oxide particle via a sol-gel approach. They synthesized a metal alkoxide derivative by the substitution of an alkoxide group with a functionalized pentane-2,4-dione derivative having an initiating group for ATRP, and synthesized surface-modified amorphous metal oxide particles of titanium, zirconium, tantalum, yttrium, and vanadium using corresponding metal alkoxides in a microemulsion-based sol-gel process [103].

Organic particles were also used as a substrate for surface-initiated ATRP. Charleux et al. synthesized a functionalized latex bearing initiation sites for ATRP on the surface by the emulsion polymerization of ST, divinyl benzene, and an initiator-holding monomer [104]. They obtained latex macroinitiators for the surface-initiated ATRP of water-soluble monomers. Stöver et al. synthesized a narrowly size-distributed polymeric microsphere with initiating sites for ATRP [105]. They first prepared a hydroxyl group-functionalized, narrowly size-distributed microsphere by precipitation copolymerization of divinyl benzene and hydroxyl methacrylate, and subsequently modified the microsphere by reaction with an initiator-holding acid bromide. Brooks et al. synthesized an initiator-functionalized PS latex for aqueous surface-initiated ATRP by the shell growth polymerization of ST and an initiator-holding monomer from a negatively charged PS seed latex [106]. They succeeded in the synthesis of a PS latex coated with an environmentally responsive polymer brush by aqueous surface-initiated ATRP of N-isopropylacrylamide [107].

These newly developed hybrid particles would have many possible applications due to their high surface functionality and unique structural feature. Here we will show some examples of application of fine particles coated with high-density polymer brushes, citing the work carried out by the authors' group. As already described, the high-density PMMA brush formed on a silicon wafer is swollen in good solvent to give a film thickness nearly equal to the full contour length of the graft chains. This surprising phenomenon led us to postulate that flexible polymers like PS and PMMA densely end-grafted on a spherical particle will also take, in solution, a highly extended chain conformation exerting interparticle interactions of an extremely long range, a long range comparable to the contour length of the grafts. Such a long range interaction, which has never been realized in a nonelectrolytic system, could lead to a two- or three-dimensionally ordered assembly of the particles. Recent studies do suggest that this in fact is the case with narrowly size-distributed AuNP and SiP [97, 108].

A narrowly size-distributed AuNP coated with a high-density PMMA brush was synthesized by surface initiated ATRP [108]. A solution of the AuNP-PMMA hybrid was spread onto the surface of purified water in a circular Langmuir trough [197] equipped with a circular compression barrier. The formed surface films of AuNP-PMMA hybrids were transferred onto a carbon-coated copper transmission electron microscopic (TEM) grid mounted on a glass plate by the Langmuir–Blodgett deposition (vertical lifting-up) technique at a constant surface pressure of 25 mN m^{-1}. Figures 22a and b show the TEM images of the transferred films. The TEM images show that the AuNP cores visible as dark circles are uniformly dispersed throughout the film, while the PMMA chains, which should be form-

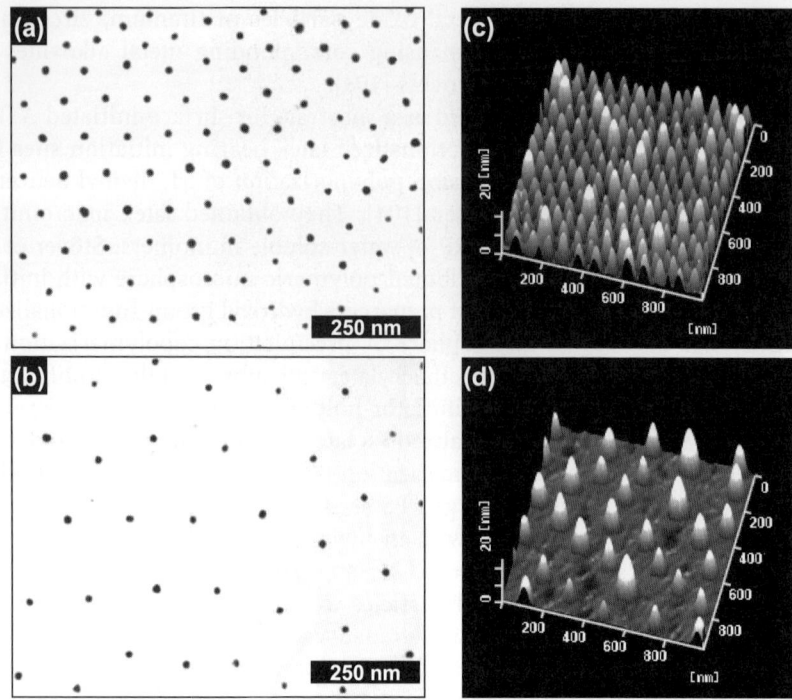

Fig. 22 TEM and AFM images of PMMA-coated AuNPs: M_ns of the graft polymers are (**a, c**) 28 000 and (**b, d**) 62 000

ing fringes surrounding the AuNP cores, are invisible due to their much lower electron density. Noteworthy are the high degree of positional order of the AuNP cores achieved in the hybrid films and the strong dependence of the interparticle distance on the chain length of PMMA grafts. The same monolayer films were transferred onto a freshly cleaved mica substrate and also observed on an AFM. The AFM images given in Figs. 22c and d exhibit protrusions whose height and thickness both increase with increasing chain length of PMMA grafts. In the AFM image in Fig. 22d, for example, the protrusions are standing on the rather flat but rough surface, which should not be the substrate (mica) surface but a PMMA matrix, as a mica surface should look perfectly smooth in this magnification. The mean nearest-neighbor center-to-center distance of the AuNP-PMMA hybrid is much larger than the diameter of the "compact core-shell model", which consists of the AuNP core and the PMMA shell of the bulk density. It is as large as about half the diameter of the "fully stretched core-shell model", which consists of the AuNP core and the shell assuming the PMMA chains radially stretched in all *trans* conformation. This means that the PMMA shell is not compact but has an extremely large extension parallel to the surface, thus producing the strong hindrance against compression. As a result, the AuNPs form a two-dimensional array

Fig. 23 Confocal laser scanning microscopic image of rhodamine-labeled SiP coated with PMMA brush: The diameter of silica particle core is 230 nm, and the M_n of the graft polymer is 256 000

with a high degree of structural order and a wide controllability of the interparticle distance.

Owing to the simplicity and versatility of surface-initiated ATRP, the above-mentioned AuNP work may be extended to other particles for their two- or three-dimensionally ordered assemblies with a wide controllability of lattice parameters. In fact, a dispersion of monodisperse SiPs coated with high-density PMMA brushes showed an iridescent color, in organic solvents (e.g., toluene), suggesting the formation of a colloidal crystal [108]. To clarify this phenomenon, the direct observation of the concentrated dispersion of a rhodamine-labeled SiP coated with a high-density polymer brush was carried out by confocal laser scanning microscopy. As shown in Fig. 23, the experiment revealed that the hybrid particles formed a wide range of three-dimensional array with a periodic structure. This will open up a new route to the fabrication of colloidal crystals.

5.2
Application as Novel Biointerface

Another attractive application of polymer brushes is directed toward a biointerface to tune the interaction of solid surfaces with biologically important materials such as proteins and biological cells. For example, it is important to prevent surface adsorption of proteins through nonspecific interactions, because the adsorbed protein often triggers a bio-fouling, e.g., the deposition of biological cells, bacteria and so on. In an effort to understand the process of protein adsorption, the interaction between proteins and brush surfaces has been modeled: like the interaction with particles, the interaction with proteins is simplified into three generic modes. One is the primary adsorption, in which a protein (or modeled particle) diffuses into the brush and adsorbs on the surface. The secondary adsorption is the adsorption of a protein at the outermost (free) surface of swollen brushes. The last one is the tertiary adsorption, which is caused by the interaction of protein with the polymer segments within the brush. One of the earliest models for the interaction of polymer brushes with particles was presented by Jeon et al. [198, 199], who assumed that a particle had no specific attractive interaction with the polymer segments, and that the polymer segment density was constant in the brush. Later, Halperin extended the systems of Jeon et al. to derive simple scaling relationships for the interaction between polymer brushes and proteins [200]. As shown in Fig. 24, the protein was treated as a dense, rigid object with a nonadsorbing surface like a structureless colloidal particle. For relatively small proteins, primary adsorption induced by a short range attraction at the surface is important and will be repressed by an increase of σ. For larger or rod-like proteins, which hardly diffuse into the brush and cannot approach the substrate surface by a steric or osmotic repulsion between the brush and protein, the secondary adsorption is predominated by van der Waals attraction with the outermost (free) surface of the brush. This mode is proposed to be suppressed by increasing the brush thickness.

Surfaces grafted with poly(ethylene glycol) (PEO) have been most extensively studied. A detailed review of this topic was recently given in [139]. To prepare PEO surfaces, various strategies were applied: physical adsorption of

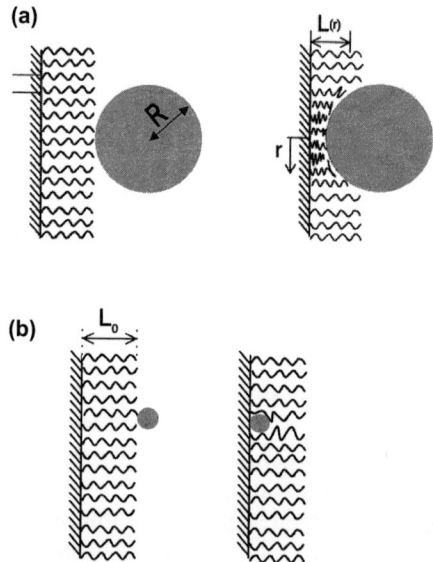

Fig. 24 **a** Primary adsorption of small proteins and **b** Secondary adsorption of large proteins. Reproduced with permission from [200] (Copyright 1999 American Chemical Society)

block copolymers or comb-like polymers and chemical grafting by, mostly, the grafting-to method. In the previously studied systems, structural parameters such as graft density are not always clear, but those brushes are presumed to range from a dilute to semi-dilute regime. Well-defined high-density polymer brushes achievable by surface-initiated LRP would produce new surfaces with (i) a size exclusion limit set for a very low molecular weight, (ii) a strong repulsion force against compression, (iii) effectively tunable properties of the outermost (free) surface by introducing functional moieties or short block segments at the end of graft chains, (iv) feasible grafting of functional polymers including hydrophilic polymers and polyelectrolytes, and (v) precise design of a variety of chain architectures such as hyperbranched and cross-linked graft chains. Ma et al. reported "nonfouling" polymer brushes synthesized by the surface-initiated ATRP of OEGMA on a Au surface coated with a SAM of a bromo-ester alkanethiol [120]. These polymer brushes exhibited no detectable adsorption of proteins, and were cell-resistant for up to a month under typical cell culture conditions. The patterned brush surfaces were also fabricated by combining the microcontact printing method and AFM dip-pen nanolithography. Spatially selective adsorption of a cell-adhesive protein, fibronectin, produced a cellular pattern after incubation with fibroblast cells.

6
Conclusion and Prospect

The successful application of LRP has enabled us to control the graft polymerization initiated from solid surfaces and to graft a variety of well-defined polymers with an exceptionally high graft density. The high graft density was demonstrated to make graft chains stretch to a high extent in the dry state, and almost to the full length of the graft chain in good solvent, providing properties, in both dry and swollen states, quite different and unpredictable from those of the previously studied semi-dilute polymer brushes. Current studies on surface-initiated LRP and thereby formed high-density polymer brushes are bringing about a breakthrough in the science and technology of polymer brushes.

Future efforts will be successively directed to the synthesis of more sophisticated high-density polymer brushes as well as better understanding of their structure and properties as a source of a new class of materials. From the synthetic viewpoint, further kinetic studies on surface-initiated LRP will still be needed for quantitative understanding of involved elementary reactions (in particular, activation/deactivation and termination processes), the knowledge of which would meet an increasing demand for elaborateness, novelty, and versatility of the brush architecture and hopefully bring about further increase in graft density and maximum chain length. As a post-polymerization process, the chemical modification of free ends of polymer brushes will play an increasingly important role in the brush synthesis, since the end-groups are localized at the outermost surface and hence effectively tune surface properties. This is one of the key aspects specific to high-density polymer brushes.

Another point is related to the basic science of polymer brushes. A semi-dilute brush has been intensively studied both experimentally and theoretically, establishing the brush science in a semi-dilute regime. Beyond this graft-density regime, surface-initiated LRP has realized a high-density (or dense) brush as a new category. Several important features to be addressed include static properties such as segmental density profile and segmental orientation as well as dynamic properties related to tribology. Experimental as well as theoretical and simulation approaches would be helpful for a new brush science.

These synthetic and structure/properties studies will find many potential applications, e.g., a variety of molecular devices based on unique molecular recognition/organization and tribologic properties on the brush surface. Examples are self-assembled hybrid fine particles, biointerfaces, high-performance monolithic materials, and sensor devices. Some applications require more sophisticated control in surface morphology, which may be achieved by the synchronization of lithographic patterning in a micrometer scale and self-assembling of mixed-homopolymer and block-copolymer brushes in a nanometer scale. Among a variety of functional polymer brushes fabricable by surface-initiated LRP, a high-density polyelectrolyte brush is one of the most important targets not only in practical application but also in brush science as a charged brush in a high-density regime. Systematic studies are believed to open up a new route to the production of new surfaces and new hybrid materials.

References

1. Napper DH (1983) Polymeric Stabilization of Colloidal Dispersions. Academic Press, London
2. Raphael E, Degennes PG (1992) J Phys Chem 96:4002
3. Klein J (1996) Annu Rev Mater Sci 26:581
4. Klein J, Kumacheva E (1995) Science 269:816
5. Parnas RS, Cohen Y (1994) Rheol Acta 33:485
6. Israelachvili JN (1992) Intermolecular and Surface Forces, 2nd Ed. Academic Press, London
7. Halperin A, Tirrell M, Lodge TP (1992) Adv Polym Sci 100:31
8. Kawaguchi M, Takahashi A (1992) Adv Colloid Interface Sci 37:219
9. Taunton HJ, Toprakcioglu C, Fetters LJ, Klein J (1990) Macromolecules 23:571
10. Courvoisier A, Isel F, Francois J, Maaloum M (1998) Langmuir 14:3727
11. Satija SK, Majkrzak CF, Russell TP, Sinha SK, Sirota EB, Hughes GJ (1990) Macromolecules 23:3860
12. Levicky R, Koneripalli N, Tirrell M, Satija SK (1998) Macromolecules 31:3731
13. Cosgrove T, Heath TG, Phipps JS, Richardson RM (1991) Macromolecules 24:94

14. Field JB, Toprakcioglu C, Ball RC, Stanley HB, Dai L, Barford W, Penfold J, Smith G, Hamilton W (1992) Macromolecules 25:434
15. Anastassopoulos DL, Vradis AA, Toprakcioglu C, Smith GS, Dai L (1998) Macromolecules 31:9369
16. Hadziioannou G, Patel S, Granick S, Tirrell M (1986) J Am Chem Soc 108:2869
17. Watanabe H, Tirrell M (1993) Macromolecules 26:6455
18. Ansarifar MA, Luckham PF (1988) Polymer 29:329
19. Kelley TW, Schorr PA, Johnson KD, Tirrell M, Frisbie CD (1998) Macromolecules 31:4297
20. Oshea SJ, Welland ME, Rayment T (1993) Langmuir 9:1826
21. Shim DFK, Cates ME (1989) J Phys (Paris) 50:3535
22. Lai PY, Halperin A (1991) Macromolecules 24:4981
23. Mansky P, Liu Y, Huang E, Russell TP, Hawker C (1997) Science 275:1458
24. Tsubokawa N, Maruyama K, Sone Y, Shimomura M (1989) Polym J 21:475
25. Boven G, Oosterling MLCM, Challa G, Schouten AJ (1990) Polymer 31:2377
26. Spange S, Heublein G, Simon F (1991) J Macromol Sci (Chem) A28:373
27. Tsubokawa N, Koshiba M (1997) J Macromol Sci Pure Appl Chem A34:2509
28. Prucker O, Rühe J (1998) Macromolecules 31:592
29. Prucker O, Rühe J (1998) Macromolecules 31:602
30. Edmondson S, Huck WTS (2004) J Mater Chem 14:730
31. Hawker CJ (1997) Acc Chem Res 30:373
32. Patten TE, Matyjaszewski K (1998) Adv Mater 10:901
33. Kamigaito M, Ando T, Sawamoto M (2001) Chem Rev 101:3689
34. Davis KA, Matyjaszewski K (2002) Adv Polym Sci 159:1
35. Matyjaszewski K (ed) (1998) Controlled radical polymerization. ACS Symposium Series 685. ACS, Washington, DC
36. Matyjaszewski K (ed) (2000) Controlled/living radical polymerization. ACS Symposium Series 768. ACS, Washington, DC
37. Matyjaszewski K (ed) (2003) Advances in controlled/living radical polymerization. ACS Symposium Series 854. ACS, Washington, DC
38. Matyjaszewski K (ed) (2002) Handbook of radical polymerization. Wiley, New York
39. Goto A, Fukuda T (2004) Prog Polym Sci 29:329
40. Uzulina I, Gaillard N, Guyot A, Claverie K (2003) Comptes Rendus Chimie 6:1375
41. Studer A (2004) Chem Soc Rev 33:267
42. Edmondson S, Osborne VL, Huck WTS (2004) Chem Soc Rev 33:14
43. McCormack CL, Lowe AB (2004) Acc Chem Res 37:312
44. Pyun J, Kowalewski T, Matyjaszewski K (2003) Macromol Rapid Commun 24:1043
45. Moad G, Mayadunne RTA, Rizzardo E, Skidmore M, Thang SH (2003) Makromol Chem 192:1
46. Madruga EL (2002) Prog Polym Sci 27:1879
47. Cunningham MF (2002) Prog Polym Sci 27:1039
48. Qiu J, Charleux B, Matyjaszewski K (2001) Prog Polym Sci 26:2083
49. Coessens V, Pintauer T, Matyjaszewski K (2001) Prog Polym Sci 26:337
50. Hawker CJ, Bosman AW, Harth E (2001) Chem Rev 101:3661
51. Fischer H (2001) Chem Rev 101:3581
52. Matyjaszewski K, Xia JH (2001) Chem Rev 101:2921
53. Pyun J, Matyjaszewski K (2001) Chem Mater 13:3436
54. Asandei AD, Percec V (2001) J Polym Sci Part A Polym Chem 39:3392
55. Otsu T (2000) J Polym Sci Part A Polym Chem 38:2121
56. Fukuda T, Goto A, Ohno K (2000) Macromol Rapid Commun 21:151

57. Solomon EH, Rizzardo E, Cacioli P (1985) Eur Pat Appl EP135280
58. Georges MK, Veregin RPN, Kazmaier PM, Hamer GK (1993) Macromolecules 26:2987
59. Benoit D, Chaplinski V, Braslau R, Hawker CJ (1999) J Am Chem Soc 121:3904
60. Benoit D, Grimaldi S, Robin S, Finet JP, Tordo P, Gnanou Y (2000) J Am Chem Soc 122:5929
61. Yutani Y, Tatemoto M (1992) Eur Pat Appl EP489370A1
62. Kato M, Kamigaito M, Sawamoto M, Higashimura T (1994) Polym Prepr Jpn 43:225
63. Matyjaszewski K, Gaynor S, Wang JS (1995) Macromolecules 28:2093
64. Kato M, Kamigaito M, Sawamoto M, Higashimura T (1995) Macromolecules 28:1721
65. Wang JS, Matyjaszewski K (1995) J Am Chem Soc 117:5614
66. Otsu T, Yoshida M (1982) Macromol Chem, Rapid Commun 3:127
67. Krstina J, Moad G, Rizzardo E, Winzor CL, Berge CT, Fryd M (1995) Macromolecules 28:5381
68. Chiefari J, Chong YK, Ercole F, Krstina J, Jeffery J, Le TPT, Mayadunne RTA, Meijs GF, Moad CL, Moad G, Rizzardo E, Thang SH (1998) Macromolecules 31:5559
69. Yamago S, Iida K, Yoshida J (2002) J Am Chem Soc 124:2874
70. Wayland BB, Poszmik G, Mukerjee SL, Fryd M (1994) J Am Chem Soc 116:7943
71. Ejaz M, Yamamoto S, Ohno K, Tsujii Y, Fukuda T (1998) Macromolecules 31:5934
72. Huang WX, Wirth MJ (1999) Macromolecules 32:1694
73. Matyjaszewski K, Miller PJ, Shukla N, Immaraporn B, Gelman A, Luokala BB, Siclovan TM, Kickelbick G, Vallant T, Hoffmann H, Pakula T (1999) Macromolecules 32:8716
74. Husseman M, Malmstrom EE, McNamara M, Mate M, Mecerreyes D, Benoit DG, Hedrick JL, Mansky P, Huang E, Russell TP, Hawker CJ (1999) Macromolecules 32:1424
75. Xiao DQ, Wirth MJ (2002) Macromolecules 35:2919
76. Jeyaprakash JD, Samuel S, Dhamodharan R, Rühe J (2002) Macromol Rapid Commun 23:277
77. Ramakrishnan A, Dhamodharan R, Rühe J (2002) Macromol Rapid Commun 23:612
78. Parvole J, Laruelle G, Guimon C, Francois J, Billon L (2003) Macromol Rapid Commun 24:1074
79. Yu WH, Kang ET, Neoh KG, Zhu SP (2003) J Phys Chem B 107:10198
80. Kim JB, Bruening ML, Baker GL (2000) J Am Chem Soc 122:7616
81. Kim JB, Huang WX, Miller MD, Baker GL, Bruening ML (2003) J Polym Sci Part A Polym Chem 41:386
82. Huang WX, Kim JB, Bruening ML, Baker GL (2002) Macromolecules 35:1175
83. Jones DM, Brown AA, Huck WTS (2002) Langmuir 18:1265
84. Desai SM, Solanky SS, Mandale AB, Rathore K, Singh RP (2003) Polymer 44:7645
85. von Werne T, Patten TE (1999) J Am Chem Soc 121:7409
86. von Werne T, Patten TE (2001) J Am Chem Soc 123:7497
87. Pyun J, Matyjaszewski K, Kowalewski T, Savin D, Patterson G, Kickelbick G, Huesing N (2001) J Am Chem Soc 123:9445
88. Pyun J, Jia SJ, Kowalewski T, Patterson GD, Matyjaszewski K (2003) Macromolecules 36:5094
89. Ohno K, Morinaga T, Koh K, Tsujii Y, Fukuda T (2005) Macromolecules 38:2137
90. Carrot G, Diamanti S, Manuszak M, Charleux B, Vairon IP (2001) J Polym Sci Part A Polym Chem 39:4294
91. Mori H, Seng DC, Zhang MF, Muller AHE (2002) Langmuir 18:3682
92. Chen XY, Armes SR (2003) Adv Mater 15:1558
93. Chen XY, Armes SP, Greaves SJ, Watts JF (2004) Langmuir 20:587

94. Nuss S, Bottcher H, Wurm H, Hallensleben ML (2001) Angew Chem Int Ed 40:4016
95. Mandal TK, Fleming MS, Walt DR (2002) Nano Lett 2:3
96. Ohno K, Koh K, Tsujii Y, Fukuda T (2002) Macromolecules 35:8989
97. Ohno K, Koh K, Tsujii Y, Fukuda T (2003) Angew Chem Int Ed 42:2751
98. Wang Y, Teng XW, Wang JS, Yang H (2003) Nano Lett 3:789
99. Marutani E, Yamamoto S, Ninjbadgar T, Tsujii Y, Fukuda T, Takano M (2004) Polymer 45:2231
100. Farmer SC, Patten TE (2001) Chem Mater 13:3920
101. Gu B, Sen A (2002) Macromolecules 35:8913
102. Liu TQ, Jia S, Kowalewski T, Matyjaszewski K, Casado-Portilla R, Belmont J (2003) Langmuir 19:6342
103. Holzinger D, Kickelbick G (2003) Chem Mater 15:4944
104. Guerrini MM, Charleux B, Vairon JP (2000) Macromol Rapid Commun 21:669
105. Zheng GD, Stover HDH (2002) Macromolecules 35:7612
106. Jayachandran KN, Takacs-Cox A, Brooks DE (2002) Macromolecules 35:4247
107. Kizhakkedathu JN, Norris-Jones R, Brooks DE (2004) Macromolecules 37:734
108. Ohno K, Morinaga T, Takeno S, Tsujii Y, Fukuda T (2006) Macromolecules (in press)
109. Huang XY, Wirth MJ (1997) Anal Chem 69:4577
110. Huang XY, Doneski LJ, Wirth MJ (1998) Anal Chem 70:4023
111. Habaue S, Ikeshima O, Ajiro H, Okamoto Y (2001) Polym J 33:902
112. Feldmann A, Claussnitzer U, Otto M (2004) J Chromatogr B Anal Technol Biomed Life Sci 803:149
113. Ejaz M, Tsujii Y, Fukuda T (2001) Polymer 42:6811
114. Blomberg S, Ostberg S, Harth E, Bosman AW, Van Horn B, Hawker CJ (2002) J Polym Sci Part A Polym Chem 40:1309
115. Ejaz M (2001) Doctoral Thesis. Kyoto University
116. Yamamoto S, Ejaz M, Tsujii Y, Matsumoto M, Fukuda T (2000) Macromolecules 33:5602
117. Yamamoto S, Ejaz M, Tsujii Y, Fukuda T (2000) Macromolecules 33:5608
118. Fukuda T (2004) J Polym Sci Part A Polym Chem 42:4743
119. Yamamoto S, Tsujii Y, Fukuda T (2000) Macromolecules 33:5995
120. Ma HW, Hyun JH, Stiller P, Chilkoti A (2004) Adv Mater 16:338
121. Milchev A, Wittmer JP, Landau DP (2000) J Chem Phys 112:1606
122. Devaux C, Chapel JP, Beyou E, Chaumont P (2002) Eur Phys J E 7:345
123. Beyou E, Humbert J, Chaumont P (2003) e-Polymer 020
124. Bartholome C, Beyou E, Bourgeat-Lami E, Chaumont P, Zydowicz N (2003) Macromolecules 36:7946
125. Parvole J, Billon L, Montfort JP (2002) Polym Int 51:1111
126. Matsumoto A (2003) Polym J 35:93
127. Baum M, Brittain WJ (2002) Macromolecules 35:610
128. Zhai GQ, Yu WH, Kang ET, Neoh KG, Huang CC, Liaw DJ (2004) Ind Eng Chem Res 43:1673
129. Tsujii Y, Ejaz M, Sato K, Goto A, Fukuda T (2001) Macromolecules 34:8872
130. Derjaguin BV (1934) Kolloid Zeits 69:155
131. Milner ST, Witten TA, Cates ME (1988) Macromolecules 21:2610
132. Milner ST, Witten TA, Cates ME (1989) Macromolecules 22:853
133. Wu T, Efimenko K, Genzer J (2002) J Am Chem Soc 124:9394
134. Wu T, Efimenko K, Vlcek P, Subr V, Genzer J (2003) Macromolecules 36:2448
135. Pakula T (1999) Makromol Chem 139:49

136. Balamurugan S, Mendez S, Balamurugan SS, O'Brien MJ, Lopez GP (2003) Langmuir 19:2545
137. Goodman D, Kizhakkedathu JN, Brooks DE (2004) Langmuir 20:2333
138. de Gennes PG (1987) Adv Colloid Interface Sci 27:189
139. Currie EPK, Norde W, Stuart MAC (2003) Adv Colloid Interface Sci 100:205
140. Ishizuka N, Minakuchi H, Nkanishi K, Hirao K, Tanaka N (2001) Colloids Surf A 187–188:273
141. Forrest JA, Dalnoki-Veress K (2001) Adv Colloid Interface Sci 94:167
142. Fryer DS, Nealey PF, de Pablo JJ (2000) Macromolecules 33:6439
143. Fryer DS, Peters RD, Kim EJ, Tomaszewski JE, de Pablo JJ, Nealey PF, White CC, Wu WL (2001) Macromolecules 34:5627
144. Jones RAL (1999) Curr Opin Colloid Interface Sci 4:153
145. Keddie JL, Jones RAL (1995) Isr J Chem 35:21
146. Prucker O, Christian S, Bock H, Rühe J, Frank CW, Knoll W (1998) Macromol Chem Phys 199:1435
147. Yamamoto S, Tsujii Y, Fukuda T (2002) Macromolecules 35:6077
148. Tanaka K, Kojio K, Kimura R, Takahara A, Kajiyama T (2003) Polym J 35:44
149. Savin DA, Pyun J, Patterson GD, Kowalewski T, Matyjaszewski K (2002) J Polym Sci Part B Polym Phys 40:2667
150. Urayama K, Yamamoto S, Tsujii Y, Fukuda T, Neher D (2002) Macromolecules 35:9459
151. Urayama K, Kircher O, Bohmer R, Neher D (1999) J Appl Phys 86:6367
152. Urayama K, Tsuji M, Neher D (2000) Macromolecules 33:8269
153. Winkelhahn HJ, Pakula T, Neher D (1996) Macromolecules 29:6865
154. Yamamoto S, Tsujii Y, Fukuda T, Torikai N, Takeda M (2001–2002) KENS Report 14:204
155. Aubouy M, Fredrickson GH, Pincus P, Raphael E (1995) Macromolecules 28:2979
156. Bergbreiter DE, Tao CL (2000) J Polym Sci Part A Polym Chem 38:3944
157. Huang WX, Baker GL, Bruening ML (2001) Angew Chem Int Ed 40:1510
158. Mori H, Boker A, Krausch G, Muller AHE (2001) Macromolecules 34:6871
159. Wang JY, Chen W, Liu AH, Lu G, Zhang G, Zhang JH, Yang B (2002) J Am Chem Soc 124:13358
160. Ejaz M, Ohno K, Tsujii Y, Fukuda T (2000) Macromolecules 33:2870
161. Rühe J, Ballauff M, Biesalski M, Dziezok P, Grohn F, Johannsmann D, Houbenov N, Hugenberg N, Konradi R, Minko S, Motornov M, Netz RR, Schmidt M, Seidel C, Stamm M, Stephan T, Usov D, Zhang HN (2004) Adv Polym Sci 165:79
162. Abe T, Hayashi S, Higashi N, Niwa M, Kurihara K (2000) Colloids Surf A 169:351
163. Tsujii Y, Hirose Y, Ejaz M, Fukuda T, Ishidoya M (2002) Polym Prepr (Am Chem Soc, Div polym Chem) 43:317
164. Prucker O, Schimmel M, Tovar G, Knoll W, Rühe J (1998) Adv Mater 10:1073
165. De Boer B, Simon HK, Werts MPL, van der Vegte EW, Hadziioannou G (2000) Macromolecules 33:349
166. Hou SF, Li ZC, Li QG, Liu ZF (2004) Appl Surf Sci 222:338
167. Hua F, Shi J, Lvov Y, Cui T (2002) Nano Lett 2:1219
168. Husemann M, Morrison M, Benoit D, Frommer KJ, Mate CM, Hinsberg WD, Hedrick JL, Hawker CJ (2000) J Am Chem Soc 122:1844
169. Jones DM, Huck WTS (2001) Adv Mater 13:1256
170. Jones DM, Smith JR, Huck WTS, Alexander C (2002) Adv Mater 14:1130
171. Maeng IS, Park JW (2003) Langmuir 19:4519
172. Schmelmer U, Jordan R, Geyer W, Eck W, Golzhauser A, Grunze M, Ulman A (2003) Angew Chem Int Ed 42:559

173. Shah RR, Merreceyes D, Husemann M, Rees I, Abbott NL, Hawker CJ, Hedrick JL (2000) Macromolecules 33:597
174. Zhou F, Liu WM, Hao JC, Xu T, Chen M, Xue QJ (2003) Adv Func Mater 13:938
175. Ejaz M, Yamamoto S, Tsujii Y, Fukuda T (2002) Macromolecules 35:1412
176. Tsujii Y, Ejaz M, Yamamoto S, Fukuda T, Shigeto K, Mibu K, Shinjo T (2002) Polymer 43:3837
177. Boyes SG, Akgun B, Brittain WJ, Foster MD (2003) Macromolecules 36:9539
178. Boyes SG, Brittain WJ, Weng X, Cheng SZD (2002) Macromolecules 35:4960
179. Carlmark A, Malmstrom EE (2003) Biomacromolecules 4:1740
180. Kim JB, Huang WX, Bruening ML, Baker GL (2002) Macromolecules 35:5410
181. Kong XX, Kawai T, Abe J, Iyoda T (2001) Macromolecules 34:1837
182. Osborne VL, Jones DM, Huck WTS (2002) Chem Commun 17:1838
183. Sedjo RA, Mirous BK, Brittain WJ (2000) Macromolecules 33:1492
184. Zhao B, Brittain WJ (2000) Macromolecules 33:8813
185. Zhao B, Brittain WJ, Zhou WS, Cheng SZD (2000) J Am Chem Soc 122:2407
186. Zhao B, Brittain WJ, Zhou WS, Cheng SZD (2000) Macromolecules 33:8821
187. Marko JF, Witten TA (1991) Phys Rev Lett 66:1541
188. Brown G, Chakrabarti A, Marko JF (1994) Europhys Lett 25:239
189. Soga KG, Zuckermann MJ, Guo H (1996) Macromolecules 29:1998
190. Muller M (2002) Phys Rev E 65:
191. Minko S, Muller M, Usov D, Scholl A, Froeck C, Stamm M (2002) Phys Rev Lett 88:035502
192. Sidorenko A, Minko S, Schenk-Meuser K, Duschner H, Stamm M (1999) Langmuir 15:8349
193. Zhao B (2003) Polymer 44:4079
194. Zhao B, He T (2003) Macromolecules 36:8599
195. Ejaz M, Ohno K, Tsujii Y, Fukuda T (2003) Polym Prepr (Am Chem Soc, Div polym Chem) 44:532
196. Zhao B, Haasch RT, MacLaren S (2004) J Am Chem Soc 126:6124
197. Matsumoto M, Tsujii Y, Nakamura KI, Yoshimoto T (1996) Thin Solid Films 280:238
198. Jeon SI, Andrade JD (1991) J Colloid Interface Sci 142:159
199. Jeon SI, Lee JH, Andrade JD, Degennes PG (1991) J Colloid Interface Sci 142:149
200. Halperin A (1999) Langmuir 15:2525
201. Bijsterbosch HD, Dehaan VO, Degraaf AW, Mellema M, Leermakers FAM, Stuart MAC, Vanwell AA (1995) Langmuir 11:4467

Photoinitiated Synthesis of Grafted Polymers

Daniel J. Dyer

Department of Chemistry, Southern Illinois University, Carbondale, Il 62901-4409, USA
ddyer@chem.siu.edu

1	Introduction	48
2	Photoinitiation Strategies	49
3	Substrates	51
3.1	Organic Polymers	51
3.1.1	Rubber	51
3.1.2	Polyolefins	52
3.1.3	Miscellaneous Organic Substrates	53
3.2	Silicon Oxide	53
3.3	Gold	57
4	Mixed Polymer Brushes	61
5	Future Directions	63
	References	63

Abstract We review various methods for the photochemical grafting of organic polymers to various substrates including, organic films, membranes, planar gold, silicon wafers, glass, silica gel, silica nanoparticles, and polydimethylsiloxane micro-channels. An emphasis is placed on photoinitiated synthesis of polymer brushes from planar gold and silicon.

Keywords AIBN · Iniferter · Microfluidics · Polymer brush · Surface initiated polymerization

Abbreviations

AFM   atomic force microscopy
AIBN   2,2′-azobisisobutyronitrile
ATR   attenuated total reflection
ATRP   atom transfer radical polymerization
BP   benzophenone
GF   grafting from
GT   grafting to
IR   infrared
M_n number-average molecular weight
M_w weight-average molecular weight
MMA   methyl methacrylate
NIPAAm *N*-isopropylacrylamide

NMR	nuclear magnetic resonance
PAN	polyacrylonitrile
PDMS	polydimethylsiloxane
PE	polyethylene
PET	polyethylene terephthalate
PDI	polydispersity index
PMMA	poly(methyl methacrylate)
PS	polystyrene
QCM	quartz crystal microbalance
RAIRS	reflection-absorption infrared spectroscopy
SAM	self-assembled monolayer
SBDC	N,N-(diethylamino)dithiocarbamoylbenzyl(trimethoxy)silane
SFM	scanning force microscopy
SIP	surface-initiated polymerization
THF	tetrahydrofuran
UV	ultra-violet
XPS	X-ray photoelectron spectroscopy

1
Introduction

The grafting of polymers to substrates has been studied for over fifty years and remains an important goal in polymer science. Recent work has focused on the synthesis of so-called polymer brushes whereby the polymer chains stretch out away from the substrate or interface [1–5]. This contemporary topic is a direct descendent of earlier work on organic graft copolymers in industry and academia. Research in this area is driven by the need to control the interfacial properties of films and the compatibility of blends.

There are three primary methods for modifying a planar substrate with an organic polymer (Fig. 1). This includes spin casting and dip coating, however the polymer is merely adsorbed to the substrate and may diffuse away when the substrate is immersed into a solvent in which it is soluble. Robust films may be created by utilizing a self-assembled monolayer (SAM) in order to immobilize a reactive functionality. Thus, the polymer can be attached to the surface provided the preformed polymer possesses a functional group that is capable of bonding with the surface (e.g. a polymer containing a primary amine could form an amide bond with a carboxylic acid-terminated SAM). This approach is known as the "grafting-to" (GT) technique and has been quite successful at synthesizing robust films of 1–10 nm in thickness. However, the GT technique is limited by diffusion barriers that prevent preformed polymer from intercalating through the tethered polymer to the reactive substrate. Therefore, the GT method yields a low-density brush. In contrast, the "grafting-from" (GF) approach has been utilized to synthesize high-density polymer brushes [6]. These high-density brushes can be much thicker and range from a few nanometers to greater than a micron. The large increase in

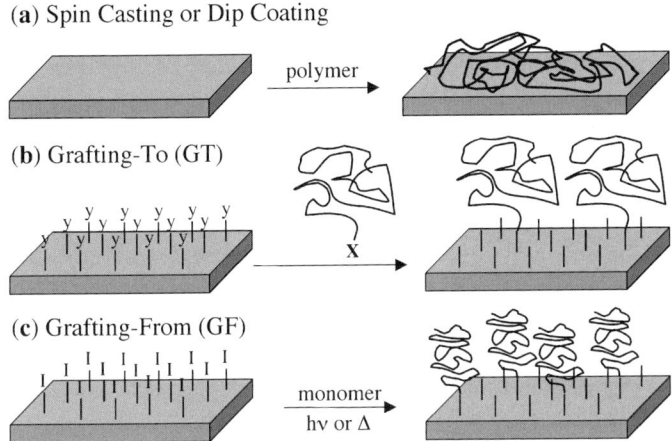

Fig. 1 Approaches for modifying a planar substrate with an organic polymer

thickness for GF films is due to much higher grafting densities compared to GT films.

Photochemical initiation has several advantages over thermal initiation. First, some functional groups are not thermally stable and so it is desirable to activate the polymerization at room temperature. This also simplifies manufacturing processes. Furthermore, most alkylthiolate SAMs are not stable above 70 °C and will begin to degrade at the temperatures required for most thermal initiations. Second, photoinitiation is usually faster than thermal initiation. Third, the initiation process may be activated at almost any temperature, this yields great flexibility when controlling the reactivity and processibility of a film. This is particularly important if one desires to initiate a polymerization in an anisotropic media, such as a liquid crystalline solvent, which may only be stable over a narrow temperature range. And most importantly, photoinitiation allows for the lithographic patterning of substrates [7, 8].

In this review we will summarize photochemical approaches for the grafting of organic polymers to various substrates. An emphasis will be placed on results from the past five years and we apologize in advance for any work that was inadvertently omitted.

2
Photoinitiation Strategies

Surface-initiated polymerization (SIP) has been carried out with a variety of initiators and Fig. 2 describes some of the most common photoinitiators [9–14]. For SIP applications, these initiators are typically modified and

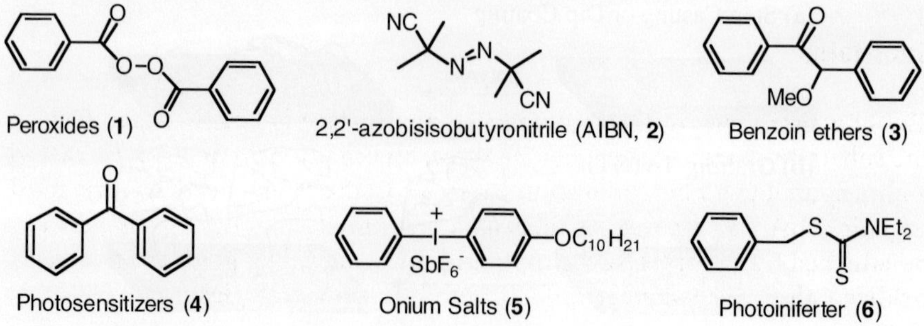

Fig. 2 Various types of photoinitiators: (1) peroxides, (2) azo compounds based on AIBN, (3) benzoin ethers, (4) triplet photosensitizers, (5) onium salts for cationic polymerization, and (6) controlled free radical polymerization with photoiniferters

covalently bonded to the substrate to yield a GF polymerization. Alternatively photosensitizers can be added to bulk solutions in order to abstract hydrogen atoms from the substrate. For example, benzophenone (BP, 4) is converted to a reactive triplet state after ultra-violet (UV) absorption, this triplet is capable of abstracting hydrogen atoms from various moieties. In particular, tertiary amines [9, 15, 16] and thiol-ene [17–19] systems have been activated with photosensitizers, but until recently have not been used for SIP. Other photoinitiators include peroxides (1) and benzoin derivatives (3), of the two, only peroxides have been used for SIP. The most common free radical photoinitiators are derivatives of 2,2′-azobisisobutyronitrile (AIBN) and these have been used by several research groups for GF polymerizations from various substrates.

Recently controlled free radical polymerizations have gained much recognition owing to their low polydispersities and "living"-like properties [20–23]. Indeed, "living" polymerizations have a tremendous advantage for SIP since it is possible to grow block copolymers and to terminate the polymerization with specific end-groups. However, most living free radical polymerizations utilize thermal initiation; for example, atom transfer radical polymerization (ATRP) may be used for SIP [24], but the rate of initiation and propagation is relatively slow compared to photoinitiation and typically leads to films that are less than 50 nm thick. Nevertheless, controlled photoinitiation strategies are under development and most are based on chain transfer agents. These so-called iniferter (*initiator-transfer-terminator*) systems utilize a reversible chain transfer process that is activated by heat or light. Compound (6) is a well-known iniferter and derivatives have been used for SIP (Sects. 3.2 and 3.3). Another interesting class of initiators that have not been used for SIP are the so-called onium salts (compound 5), which are used for photoinitiated cationic polymerization [25]. Thus, there are a rich variety of photopolymerization strategies that may be utilized and very few examples of photo-SIP have been reported to date.

3
Substrates

The chemistry for grafting polymers to any surface is largely dependent on the substrate characteristics. For instance, an organic polymer can be cast as a film, spun into a fiber, or extruded into a specific shape. Furthermore, most organic substrates are easily oxidized so that initiators or other polymers can be attached. In contrast, some inorganic substrates, like gold, are not easily oxidized; thus, necessitating different modification strategies.

3.1
Organic Polymers

The development of photografting techniques for the modification of organic polymers is largely a result of the poor adhesion and solubility properties of polyethylene (PE) and related polyolefins. In particular, there is a great desire to increase the solubility of polyolefins for improved processing. Furthermore, improved adhesion is sought for blends and to improve interactions with other substrates. The adhesion of organic dyes to polyolefins and natural fibers helped drive early developments in this field [26–28]. The synthetic methodology for graft polymerization has been thoroughly reviewed [26–28], therefore, we will briefly highlight some of this previous work and focus most of our discussions on recent results. A typical photografting process involves the deposition of a mixture of photosensitizer and polymer onto an organic substrate; this is followed by a UV curing step prior to a final cleaning step with an appropriate solvent.

3.1.1
Rubber

The earliest example of photografting is largely the result of work by Charlesby and colleagues in the early 1950s [29], who discovered that γ-irradiation could be used to cross-link PE. Oster quickly realized that ionizing radiation was less than ideal since it was damaging to the bulk film as well as to the surface. Furthermore, γ-irradiation is relatively dangerous to work with and much more expensive compared to ultra-violet light. He reasoned that UV light was sufficiently energetic that it could be used to abstract hydrogen atoms from polymer films in order to create surface bound radicals, which in turn would initiate a polymerization; thus, grafting one polymer to another. However, in order to maximize the efficiency of this process a photosensitizer was added. In his seminal work, Oster cast BP from solution onto a film of unvulcanized rubber [30]. Upon irradiation in the presence of an aqueous solution of acrylamide, he was able to graft polyacrylamide in 1–5 minutes. These early studies generated a great deal of excitement in the polymer community and ultimately led to the development of new photografting procedures.

3.1.2
Polyolefins

Shortly after the initial studies on rubber, Oster and coworkers cross-linked PE by a similar procedure. They were able to graft polymers by vapor deposition [31] from a solution of BP and monomer under UV irradiation. More specifically, they grafted styrene and acrylamide to PE films and examined a variety of photosensitizers; with benzophenone and chlorobenzophenone performing best.

Recently, Yang has synthesized thermo-sensitive track membranes in order to control porosity [32]. They grafted N-isopropylacrylamide (NIPAAm) to polyethylene terephthalate (PET) membranes with BP as a photosensitizer. An advantage of this strategy over previous work is that the photoinitiation is localized to the surface because the UV light does not penetrate into the interior of the membrane. As described in Fig. 3, the pore-covering strategy should yield a more rapid response relative to pore filling, because in the pore-covering method only a thin polymer brush is present at the surface. Whereas in the pore-filling strategy, the pores are completely filled and are expected to respond more slowly due to thermal and diffusion gradients.

Yamada and co-workers recently improved the hydrophilicity of PE by photoinduced grafting of methylacrylamide with BP as a photosensitizer [33]. They examined the adhesion strength and water adsorptivity and showed that both increase with the amount of grafted poly(methacrylamide). In addition, Bowman has photo-grafted BP to polypropylene membranes [34]. The substrate could then be activated in a second step to graft acrylic acid from a monomer solution. Surprisingly, they observed a living-like polymerization with a linear relationship between the graft polymerization rate and the monomer concentration. The resulting microfiltration membranes were made hydrophilic by this two-step process. They substantially reduced the amount of undesired homopolymer and minimized cross-linking, relative to previous methods.

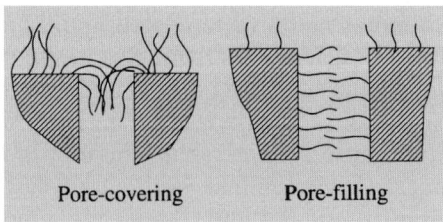

Fig. 3 Schematic of pore-covering and pore-filling strategies. (Reproduced with permission from [32]; Elsevier)

3.1.3
Miscellaneous Organic Substrates

Lee and Matsuda modified a segmented polyurethane film with a dithiocarbamate moiety and then initiated the polymerization of (ethylene glycol) methacrylate and N,N-dimethylacrylamide [35]. The resulting films were hydrophilic and were shown to inhibit adhesion of platelets. Such films could be useful as anti-coagulant substrates for devices implanted in vivo.

Recently Allbritton and Li coated polydimethylsiloxane (PDMS) microfluidic channels with BP [36]. Upon irradiation in the presence of a monomer solution, they were able to graft poly(acrylic acid) and poly(ethylene glycol) monomethoxyl acrylate to the interior walls of the channels. This is a significant achievement since the device did not require disassembly in order to modify the channel walls. The electrophoretic separation of the modified channels was different from the native channels. This technique holds particular promise for the microfluidic separations community.

3.2
Silicon Oxide

Polymers have been grafted to various SiO_x substrates including silica gel, silica nanoparticles, silicon wafers, and glass. In particular, glass and silicon wafers are easily modified with self-assembled monolayers (SAMs). The chemistry for all of these surfaces is essentially identical and varies only slightly in the cleaning procedures and characterization methods. In particular, glass is difficult to characterize by infrared (IR) spectroscopy due to the large number of Si–O absorptions in the bulk relative to the surface. Silicon wafers are more amenable to IR since only a very thin layer of SiO_x is present at the interface. Thus, the SAM absorption bands are visible via reflection-absorption infrared spectroscopy (RAIRS) or attenuated total reflection (ATR) spectroscopy. Furthermore, most IR lasers will penetrate the silicon wafer, therefore transmission mode experiments are possible. Silica gel and silica nanoparticles may be characterized with NMR, either in the solid state, or in solution if enough polymer is grafted to ensure that the particles are soluble.

Knoll and coworkers were the first to synthesize polymer brushes from SAM-coated silicon wafers via a photoinitiation strategy [37–39]. They used an AIBN type initiator that was developed by Rühe. In particular, a chlorosilane terminus was used to form a covalent bond to the native oxide surface of the silicon; this was followed by irradiation at ∼ 350 nm in the presence of styrene to yield a PS brush (Fig. 4). They were able to write patterns by irradiating through a mask to activate the surface-bound initiators. Alternatively, they synthesized a polymer film and then used deep UV ablation through masks to remove some of the polymer in the irradiated regions. By

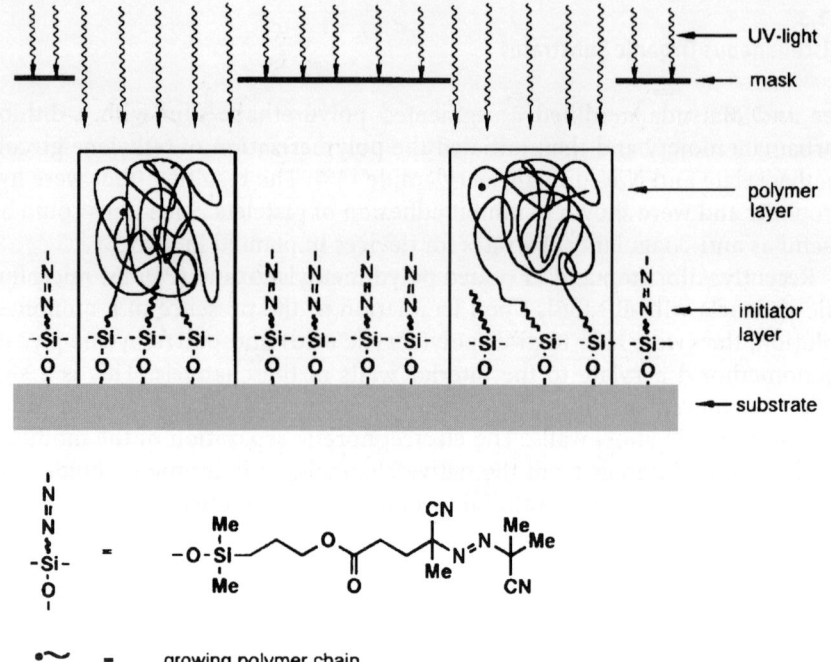

Fig. 4 Schematic representation of the procedure used to pattern silicon substrates by photopolymerization. (Reproduced with permission from [38]; Wiley)

varying the sequence of reactive steps, one can create positive or negative images of the mask on the substrate (Fig. 5). Since AIBN has a relatively low extinction coefficient, Rühe designed an aryl version with improved initiation kinetics [40]. Specifically, compound **12** in Fig. 6 exhibited a molar extinction coefficient that was three orders of magnitude larger than AIBN, and proved to be a good initiator for SIP from glass and silicon.

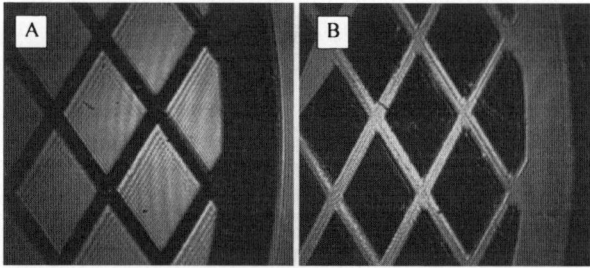

Fig. 5 Surface plasmon microscopy images of a structured ultrathin PS film prepared by photopolymerization through a mask; **A** bare silicon appears *dark*, **B** polystyrene appears *dark*. (Reproduced with permission from [38]; Wiley)

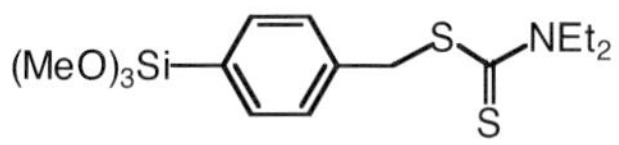

Fig. 6 Structure of known photoinitiators for silicon and gold substrates

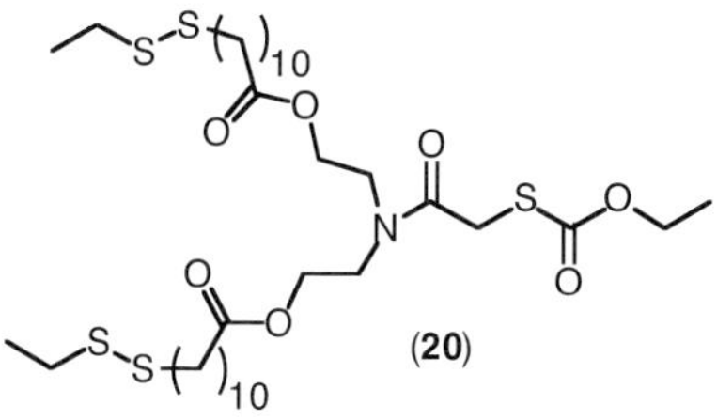

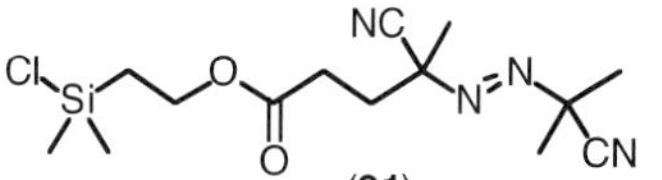

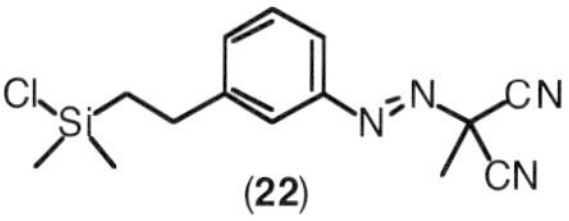

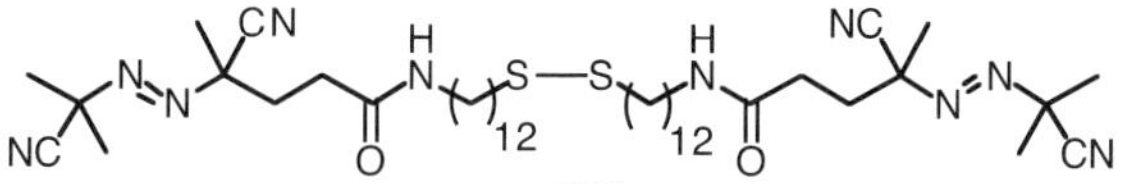

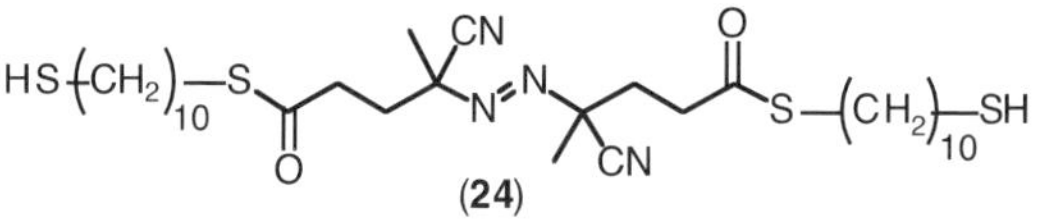

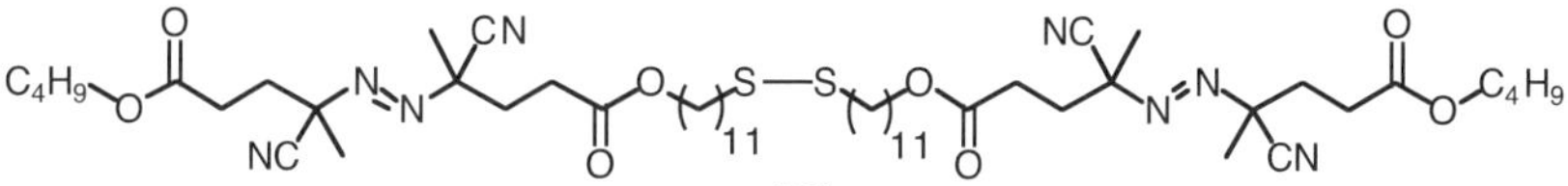

Frank and co-workers used an alternative photoactivated GT technique to immobilize polymers onto silicon [41]. They first deposited a chlorosilane SAM onto SiO_2 surfaces, which were terminated with a benzophenone moiety. They then cast films of PS and poly(ethyloxazoline) onto the substrates, which were then irradiated at ~ 340 nm. The benzophenone excited triplet would then abstract a hydrogen atom from the polymer and recombine with the activated polymer to form a robust covalent bond. They showed that the radius of gyration of the polymer had an affect on the layer thickness, up to a maximum thickness of 16 nm. Knoll and coworkers used a similar strategy to synthesize lipocopolymer substrates, in this case grafting a poly(ethylene oxide)-poly(ethyloxazoline-statethyleneimine) copolymer [42]. They then examined the adsorption of proteins onto the artificial membranes. Finally, Pinson electrochemically reduced diazonium groups prior to attachment onto iron surfaces [43]. It was then possible to irradiate in the presence of PS to graft polymer to the activated iron substrate.

Controlled free radical polymerization is an important goal in modern polymer chemistry and photochemical approaches are known. In particular, photoiniferters such as dithiocarbamoyl (**19**) have been used to modify silicon substrates. These photoiniferters provide some control over the number-average molecular weight (M_n) and polydispersity, but their greatest advantage is the ability to synthesize block copolymers. One of the earliest examples was that of Zaremski and coworkers who modified silica gel with a dithiocarbamate initiator [44]. They were able to graft up to 20% PS after 6–12 h, however the M_w/M_n was quite high (greater than 2), suggesting an uncontrolled polymerization. In contrast, Hadziioannou showed good con-

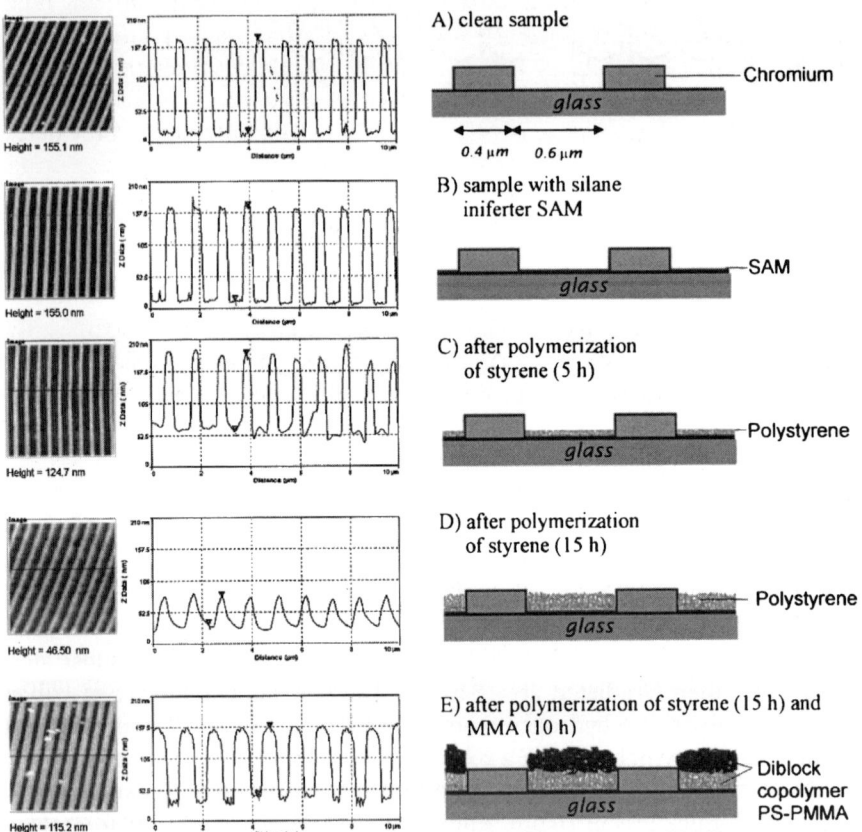

Fig. 7 Topographic images, SFM scan lines, and schematic cross-sectional representation of glass slides with chromium patterns and the formation of the polymer and copolymer layer: **A** cleaned sample, **B** sample modified with SBDC, **C** after photopolymerization of styrene for 5 h, **D** after photopolymerization of styrene for 15 h, and **E** after photopolymerization of styrene for 15 h and subsequently methyl methacrylate for 10 h. (Reproduced with permission from [45]; American Chemical Society)

trol with (compound 19) to grow a PS/poly(methyl methacrylate) (PMMA) block copolymer from a silicon wafer [45]. As Fig. 7 illustrates, they first patterned a glass substrate with chromium and then deposited the SAM onto the bare regions of SiO_x. Irradiation for 15 h in styrene produced a 100 nm PS brush. The substrate was then immersed into methyl methacrylate (MMA) and an additional 170 nm of PMMA was grown on top of the PS layer. Qui used a similar molecule to modify silica nanoparticles and then synthesize a PMMA film from the surface-bound photoiniferter [46].

3.3
Gold

In many respects gold is more convenient than silicon for characterizing polymer brush films. For instance, the RAIR spectra from planar gold are usually superior to that from silicon wafers, especially for the SAM characterization. In addition, the S – Au bond is easily cleaved by treatment in a dilute solution of iodine. This allows for the isolation of the tethered polymer and the determination of M_n and PDI. The SAMs are easily deposited from dilute solutions of thiol or disulfide precursors. The only major disadvantage to gold is the instability of the weak S – Au bond, which is unstable above 60 °C, and under UV irradiation in air.

Niwa was the first to graft a polymer film to gold by photoinitiation of a SAM [47]. He used compound 20, with a xanthate group to initiate the polymerization of methacrylic acid from a gold coated quartz crystal. They were able to monitor the growth of the polymer film with a quartz crystal microbalance (QCM) by monitoring the resonant frequency during irradiation over a 60-min period. Interestingly, the rate of polymerization was dependent on the pH and was maximized at low pH (~ 2) and essentially stopped at neutral pH. The authors attributed the decrease in rate to electrostatic repulsion of the monomer units from the growing polymer chain as both become ionized; this inhibits combination of the monomer with the reactive radical.

Rühe was studying the photoinitiated synthesis of polymer brushes from gold substrates at about the same time as Niwa. In contrast, Rühe used a disulfide initiator based on AIBN (Compound 23, Fig. 6) to polymerize styrene from gold-coated silicon wafers, and reported a brush thickness of 904 nm after 18 h UV irradiation [40]. To our knowledge this is the thickest PS brush reported to date. Thicknesses in the range of 100–400 nm are more typical for PS brushes from AIBN-type SAMs. Rühe pointed out that thiols are efficient chain transfer agents and therefore could inhibit chain propagation. This would limit the thickness of films even if high grafting densities were reached. He reported problems with an initiating system that had a short linker between the anchoring group and the azo group. When longer linkers were used (> 10-CH_2), the polymerizations performed well and were comparable to siloxane-based SAMs. Subsequent work by Dyer and Green has confirmed that the S–Au bond is stable under UV polymerization conditions, provided oxygen is excluded [48].

More recently, the Dyer group has synthesized a variety of thiolate-based AIBN-type initiators for SIP from planar gold. They reported the synthesis of PS [48, 49], PMMA [50], and polyacrylonitrile (PAN) [51] brushes from initiator (24) in Fig. 6. This bis-thiol initiator was found to bond to the gold substrate in an extended structure with a terminal thiol group at the air interface. The SAMs were disordered but proved to be versatile photoinitiators for SIP. The large concentration of sulfur in this initiator could lead to chain transfer, which might reduce the M_n of the grafted polymer. Therefore, Dyer synthesized disulfide (25) and studied the polymerization from the resulting SAMs [52]. They found that initiator (25) formed poor quality monolayers compared to (24); this significantly decreased the growth kinetics and resulted in much thinner films than with (24). In particular, SAMs of (24) yielded PS films of 190 nm after 14 h compared to 90 nm after 33 h for (25). Thus, the quality of the SAM does play a role in the overall growth kinetics and resulting film thickness. The role of chain transfer has not been quantified but it clearly does not prevent photo-SIP.

Dyer also demonstrated that a change in thickness for a brush film does not always correlate to a change in the M_n of the tethered polymer. For instance, they performed a series of photopolymerizations from SAMs of (24) and the grafted polymer was removed by treating the substrates with an iodine solution. The M_n and M_w/M_n was analyzed by GPC/MALS. Surprisingly, from 7–10 h reaction time the thickness of the PS brush increased from 70 nm to 175 nm [52]. Thus, the film thickness more than doubled over the span of 3 h, yet the M_n isolated from the grafted chains increased by only 3%. The large increase in thickness is likely due to a combination of factors including an increase in grafting density, which causes the chains to extend further out from the substrate. For this type of free radical polymerization, the M_n does not change drastically over time since the chains grow very rapidly after initiation and are followed by rapid termination. Thus, grafted chains are continuously being generated during the entire irradiation process, as long as initiator is left on the surface.

More recently SAMs containing tertiary amines have been found to be excellent photoinitiators for SIP [53]. Dyer and Yagci reasoned that a photosensitizer could be used to abstract hydrogen atoms from amine-containing SAMs. Indeed, alkyl amines have long been known to participate in photopolymerizations when activated by triplet sensitizers. As Scheme 1 illustrates, benzophenone (26) absorbs a photon and is converted to an excited singlet state, which is then converted into the reactive triplet state. This triplet state is capable of abstracting hydrogen atoms from organic species. This is particularly effective with C–H bonds that are adjacent to the nitrogen atom; radical (28) is highly reactive, whereas the ketyl radical (29) is less reactive and generally terminates. Dyer showed that disulfide (30) could form

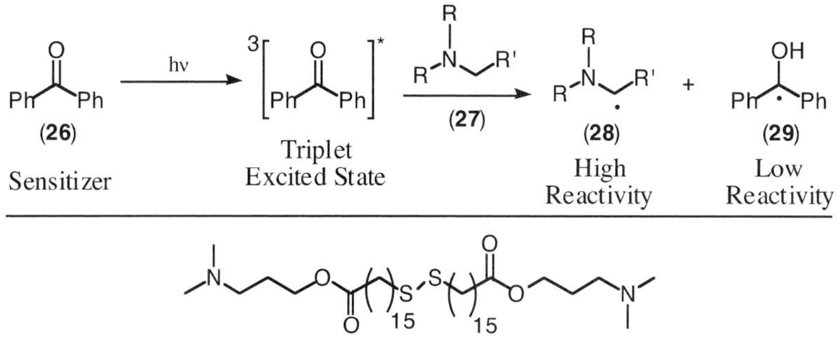

Scheme 1 A triplet photosensitizer (26) may abstract hydrogen atoms from alkyl amines (27). The resulting radical (28) is highly reactive and initiates a polymerization, while the ketyl radical (29) is less reactive. Compound (30) has been shown to form SAMs on gold and initiate polymerization [53]

densely packed SAMs on gold and that these SAMs would initiate the free radical polymerization of PS and PMMA when benzophenone was added to the monomer solution. Surprisingly, they discovered that the triplet sensitizer was not needed and the films were actually thicker and the film growth was more rapid in the absence of a photosensitizer.

Figure 8a illustrates the rapid growth of PS films from SAMs containing initiator (30). The growth kinetics are superior to anything reported to date for PS. Unfortunately, the mechanism for dimethylamino photoinitiation is unknown, but is currently under investigation in the Dyer group. The two most likely candidates involve a charge transfer complex between the terminal amine and an excited monomer molecule, or chain transfer from active polymer chains that are growing in solution. Presumably the active chains in solution result from autopolymerization, which was found with control experiments under the same reaction conditions; these active chains may abstract a hydrogen atom from the surface-bound dimethylamino groups. Thus, the polymerization would be initiated at the surface by chain transfer. While chain transfer could play a role in the early stages of the polymerization, it is unlikely that active polymer chains could rapidly diffuse to the surface as this represents the same circumstances that prevented densely packed polymer brushes via a GT approach. GT brushes are typically limited to less than 10 nm and are much slower than GF, therefore the rapid growth exhibited in Fig. 8a strongly suggests a GF polymerization.

It is interesting to compare the rapid growth from the tertiary amines with that of more traditional AIBN-type initiators. For instance, Fig. 8c represents the polymerization of styrene from SAMs of (24) on gold. The amino initiator yields much thicker films in about half the time. The rate of polymer film growth is important in manufacturing processes and it is usually desirable to minimize the UV exposure time. Partially for this reason, Rühe

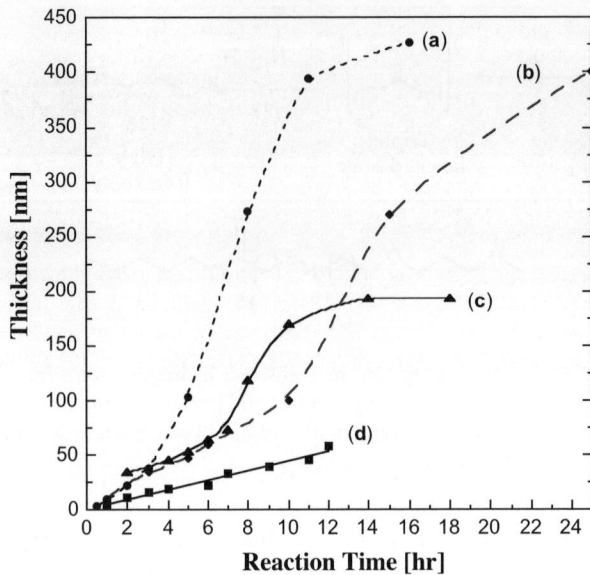

Fig. 8 Growth of polystyrene from various photoinitiating SAM-coated substrates: (a) **26** on gold [53], (b) **22** on silicon [40], (c) **24** on gold [53], and (d) **21** on silicon [54]

developed (**22**) since it is more efficient than AIBN, and as Fig. 8b illustrates, it yields a much thicker film than (**24**). However, comparisons of the rate of film growth between these initiators is complicated by significant differences in the experimental procedures. For example, Rühe's procedure typically utilized a 1 : 1 solution of styrene in toluene, whereas Dyer used neat styrene. Higher concentration should favor faster film growth. In contrast, the UV intensity in Rühe's procedure is much higher at 30 mW/cm^2 compared to 1.6 mW/cm^2 for Dyer. This would favor faster film growth for Rühe due to more rapid initiation. Since both of these factors compete against each other, it is difficult to tell which system provides the thickest films in the shortest amount of time. Furthermore, Rühe's polymerizations were from silicon substrates, whereas Dyer polymerized from gold. It is unclear if the different SAM structures play a significant role in the rate of polymer growth. One conclusion we can make is that for AIBN systems the photoinitiation leads to faster film growth than thermal initiation, as evidenced by Fig. 8d, which used initiator (**21**) and is slower than both (**24**) and (**22**), and much slower than the amino initiator (**30**).

Both photoinitiation and thermal initiation from gold generally yields very smooth surfaces. In particular, Dyer has observed values for the root-mean-square (rms) roughness on the order of ±0.2 nm for a ∼ 40 nm brush of PAN [51] and ±0.65 nm for a ∼ 140 nm PS film [52]. Figure 9 represents a typical topographic map that may be generated with an imaging ellipsometer. This provides a rapid calculation of the rms roughness along with a μm-wide view

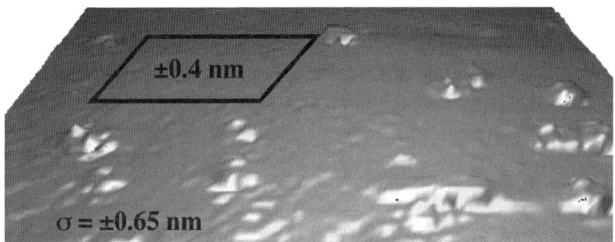

Fig. 9 A topographic thickness map generated with a scanning ellipsometer (iElli2000, Nanofilm Technologie, GmbH) of a polystyrene brush with an average thickness of 140 ± 0.6 nm over an area of 100×193 μm^2. The area within the square is very smooth, with a standard deviation of ± 4 Å. (Reproduced with permission from [52]; Wiley)

of the polymer/air interface. Dyer has reported grafting densities as high as 1.4 chains/nm^2 for PS brushes from both AIBN [48] and dimethylamino [53] initiators on gold; these are among the highest reported to date.

An alternative photo-SIP approach was described by Kang and coworkers, where they used an argon plasma to oxidize alkylthiolate SAMs on planar gold [55]. The plasma treatment oxidized carboxy-terminated SAMs to peroxide moieties. UV irradiation in the presence of acrylic acid and allylpentafluorobenzene yielded ultra-thin graft layers of ~ 6–7 nm. The poly(acrylic acid) layers were found to adsorb Fe^{3+} ions from solution. This particular photoinitiation method yields low-density polymer brush films.

4
Mixed Polymer Brushes

Recently multi-component polymer brush films have been synthesized in order to produce responsive substrates. As Fig. 10 illustrates, these so-called binary brushes contain two different polymers that may, or may not, be compatible with each other. Such a mixed brush may be synthesized by two strategies: First, a mixed monolayer may be deposited with two different initiators (Fig. 10a). For example, one initiator could be activated thermally, while the other photochemically. Thus, a low to medium density polymer brush could be synthesized by selectively activating initiator A. The substrate could then be cleaned and immersed into a second monomer solution to activate B and grow a second polymer interdispersed with the first brush.

In contrast to the mixed monolayer, a single monolayer could be used provided that the initiator was not very efficient. AIBN is an excellent choice due to its low molar absorptivity and a long half-life; therefore, the initiator will still be present after an initial period of irradiation when the first brush is deposited. The substrate could then be cleaned and reintroduced to a new monomer solution to yield a mixed brush. Both thermal and photochemi-

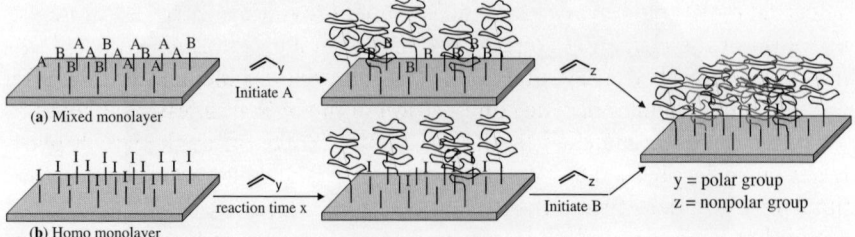

Fig. 10 Synthesis of a mixed brush may be accomplished by two strategies: (a) A free radical initiator with low efficiency is used. This guarantees that some of the initiator is left after deposition of the first brush. Immersion into a second monomer yields a mixed brush where the two polymers are intercalated between each other. (b) A mixed monolayer could be used whereby A and B may be initiated independently (e.g. one is a photoinitiator and the other is a thermal initiator)

cal activation could be used with AIBN-type initiators. Indeed, Minko and Stamm have used thermal initiation of AIBN-type SAMs to produce binary brushes [56].

Recently, Dyer used the same strategy to perform a photoinitiated synthesis of a mixed brush by using an AIBN-type initiator bound to gold [57]. Specifically, they used initiator (**24**) to modify gold substrates with a binary brush composed of PS and PMMA. As Fig. 11 describes, mixed brushes will respond to the polarity of the solvent. For example, immersion into a nonselective solvent like THF brings both components to the air/liquid interface since PS and PMMA are both soluble in THF. However, immersion into a polar solvent, such as isobutanol, will selectively bring PMMA to the air/liquid interface, while the nonpolar PS collapses into the interior of the film. In contrast, immersion into cyclohexane brings PS to the air/liquid interface and PMMA is driven to the interior. The cycle is completely reversible after immersion into a nonselective solvent like THF.

Dyer used a tandem RAIRS/X-ray photoelectron spectroscopy (XPS) technique to quantify the composition of these mixed brushes at the air/liquid interface and within the bulk film. In this case XPS was used to determine the surface composition of the film after each solvent treatment. In particular,

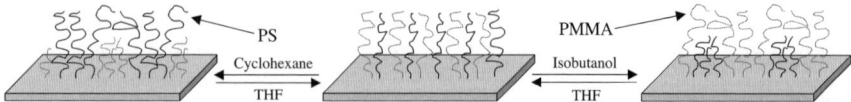

Fig. 11 Switching of a PS/PMMA brush is accomplished by immersion into various solvents. A polar solvent such as isobutanol brings PMMA to the air/liquid interface, while the PS collapses into the interior. The opposite occurs when the substrate is immersed into a nonpolar solvent such as cyclohexane. Upon immersion into a nonselective solvent, like THF, both components come to the air/liquid interface

a low take-off angle grazes the surface and provides a fairly accurate description of the elemental composition in the first 1–2 nm. As expected, the interface is enriched in PS after immersion in cyclohexane, and PMMA after immersion in isobutanol. The composition of the bulk film was confirmed by RAIRS and closely matched the XPS data at 90° take-off angle, which penetrates more deeply than at 15°. The morphology and surface roughness of the films were very sensitive to the solvent conditions. Dyer used AFM to demonstrate that these type of films exhibit lateral and perpendicular nanophase separation, which is consistent with other mixed brush studies [57].

5
Future Directions

A wide variety of initiators and photoinitiating strategies are available for the synthesis of grafted polymers. However, new initiating systems are needed, particularly to allow for the rapid synthesis of block copolymers. Current photoiniferter's are relatively slow and do not yield optimal polydispersities. Therefore, uncontrolled initiators are still needed for applications that require rapid film growth, in this context the tertiary amine systems are promising. More work is required in order to elucidate the initiating mechanism and to optimize the growth kinetics for the amino SAMs.

While photo-grafting of organic polymers is a fairly mature field, there is a need for more efficient photosensitizers. Convenient methods to graft onto natural fibers would be valuable for specialty fabrics as well as consumer products, particularly if the monomer can be delivered in the gas phase or as an aqueous solution.

The synthesis of so-called "smart materials" that respond to environmental changes will play an increasing role in interfacial science. This is particularly relevant to micro and nanofluidic applications. Modification strategies for PDMS are very important for the emerging field of microfluidic separations. Furthermore, biofluidics and tissue engineering demand precise control of the substrate properties and their subsequent interactions with biological media. Photoinitiating strategies allow for the modification of substrate properties in vitro and potentially even in vivo. Furthermore, the photoinitiated synthesis of multi-component brush films promises to be a fruitful area in the future.

References

1. Advincula RC, Brittain WJ, Caster KC, Rühe J (2004) Polymer Brushes. Wiley, Weinheim
2. Rühe J, Knoll W (2002) J Macromol Sci: Polym Rev C42:91–138
3. Zhao B, Brittain WJ (2000) Prog Polym Sci 25:677–710

4. Nagasaki Y, Kataoka K (1996) Trend Polym Sci 4:59-64
5. Milner ST (1991) Science 251:905-914
6. Edmondson S, Osborne VL, Huck WTS (2004) Chem Soc Rev 33:14-22
7. Dyer DJ (2003) Adv Funct Mater 13:667-670
8. Prucker O, Konradi R, Schimmel M, Habicht J, Park I-J, Rühe J (2004) In: Advincula RC, Brittain WJ, Caster KC, Rühe J (eds) Polymer Brushes: Synthesis, Characterization, Applications. Wiley, Weinheim, p 449-469
9. Yagci Y (2000) Macromol Symp 161:19-35
10. Fouassier JP (2000) Recent Res Devel Photochem Photobiol 4:51-74
11. Fouassier JP (2000) Recent Res Devel Polym Sci 4:131-145
12. Decker C (1998) Polym Inter 45:133-141
13. Fouassier JP, Ruhlmann D, Graff B, Wieder F (1995) Prog Org Coatings 25:169-202
14. Belfield KD, Crivello JV (eds) (2003) Photoinitiated Polymerization, ACS Symposium Series Vol 847. American Chemical Society, Washington, DC
15. Davidson RS (1993) In: Fouassier JP, Rabek JF (eds) Radiation Curing in Polymer Science and Technology: Polymerisation Mechanisms, Vol 3. Elsevier, London, p 153-176
16. Li T (1990) Polym Bull (Berlin) 24:397-404
17. Hoyle CE, Cole M, Bachemin M, Kuang W, Kalyanaraman V, Jönsson S (2003) In: Belfield KD, Crivello JV (eds) Photoinitiated Polymerization, ACS Symposium Series Vol 847. American Chemical Society, Washington, DC, p 52-64
18. Jacobine AF (1993) In: Fouassier JP, Rabek JF (eds) Radiation Curing in Polymer Science and Technology: Polymerisation Mechanisms, Vol 3. Elsevier, London, p 219-268
19. Cramer NB, Reddy SK, O'Brien AK, Bowman CN (2003) Macromolecules 36:7964-7969
20. Matyjaszewski K (ed) (2000) Controlled/Living Radical Polymerization, ACS Symposium Series, Vol 768. American Chemical Society, Washington, DC
21. Hawker CJ, Bosman AW, Harth E (2001) Chem Rev 101:3661-3688
22. Bisht HS, Chatterjee AK (2001) J Macromol Sci Polym Rev C41:139-173
23. Fischer H (2001) Chem Rev 101:3581-3610
24. Pyun J, Kowalewski T, Matyjaszewski K (2003) Macromol Rapid Commun 24:1043-1059
25. Crivello JV (2002) Designed Monomers Polym 5:141-154
26. Rånby B (1992) Makromol Chem Macromol Symp 63:55-67
27. Uyama Y, Kato K, Ikada Y (1998) Adv Polym Sci 137:1-39
28. Kato K, Uchida E, Kang E-T, Uyama Y, Ikada Y (2003) Prog Polym Sci 28:209-259
29. Charlesby A (1952) Proc Royal Soc London A215:187
30. Oster G, Shibata O (1957) J Polym Sci 26:233-234
31. Oster G, Oster GK, Moroson H (1959) J Polym Sci 24:671-681
32. Yang B, Yang W (2003) J Memb Sci 218:247-255
33. Yamada K, Takeda S, Hirata M (2003) In: Belfield KD, Crivello JV (eds) Photoinitiated Polymerization, ACS Symposium Series, Vol 847. American Chemical Society, Washington, DC, p 511-521
34. Ma H, Davis RH, Bowman CN (2000) Macromolecules 33:331-335
35. Lee HJ, Matsuda T (1999) J Biomed Mater Res 47:564-567
36. Hu S, Ren X Bachman M, Sims CE, Li GP, Allbritton NL (2004) Anal Chem 76:1865-1870
37. Tovar G, Paul S, Knoll W, Prucker O, Rühe J (1995) Supramol Sci 2:89-98
38. Prucker O, Schimmel M, Tovar G, Knoll W, Rühe J (1998) Adv Mater 10:1073-1077
39. Knoll W, Matsuzawa M, Offenhäusser A, Rühe J (1996) Israel J Chem 36:357-369
40. Prucker O, Habicht J, Park I-J, Rühe J (1999) Mater Sci Eng C 8-9:291-297

41. Prucker O, Naumann CA, Rühe J, Knoll W, Frank CW (1999) J Am Chem Soc 121:8766–8770
42. Knoll W, Frank CW, Heibel C, Naumann R, Offenhäusser A, Rühe J, Schmidt EK, Shen WW, Sinner A (2000) Rev Mol Biotechnol 74:137–158
43. Adenier A, Cabet-Deliry E, Lalot T, Pinson J, Podvorica F (2002) Chem Mater 14:4576–4585
44. Zaremski MY, Chernikova EV, Izmailov LG, Garina ES, Olenin AV (1996) Macromol Reports A33:237–242
45. de Boer B, Simon HK, Werts MPL, van der Vegte EW, Hadziioannou G (2000) Macromolecules 33:349–356
46. Bai J, Qiu K-Y, Wei Y (2003) Polym Inter 52:853–858
47. Niwa M, Date M, Higashi N (1996) Macromolecules 29:3681–3685
48. Schmidt R, Zhao T, Green J-B, Dyer DJ (2002) Langmuir 18:1281–1287
49. Paul R, Schmidt R, Feng J, Dyer DJ (2002) J Polym Sci Part A: Polym Chem 40:3284–3291
50. Dyer DJ, Feng J (2003) Polym Preprints 44(2):230–231
51. Paul R, Schmidt R, Dyer DJ (2002) Langmuir 18:8719–8723
52. Dyer DJ, Feng J, Fivelson C, Paul R, Schmidt R, Zhao T (2004) In: Advincula RC, Brittain WJ, Caster KC, Rühe J (eds) Polymer Brushes: Synthesis, Characterization, Applications. Wiley, Weinheim, p 129–150
53. Dyer DJ, Feng J, Schmidt R, Wong VN, Zhao T, Yagci Y (2004) Macromolecules 37:7072–7074
54. Prucker O, Rühe J (1998) Langmuir 14:6893–6898
55. Zhang J, Cui CQ, Lim TB, Kang E-T (2000) Macromol Chem Phys 201:1653–1661
56. Sidorenko A, Minko S, Schenk-Meuser K, Duschner H, Stamm M (1999) Langmuir 15:8349–8355
57. Feng J, Haasch RT, Dyer DJ (2004) Macromolecules 37:9525–9537

Photoiniferter-Driven Precision Surface Graft Microarchitectures for Biomedical Applications

Takehisa Matsuda

Division of Biomedical Engineering, Graduate School of Medicine, Kyushu University, Maidashi 3-1-1, 812-8582 Higashi-ku, Fukuoka, Japan
matsuda@med.kyushu-u.ac.jp

1	Introduction	68
2	**Surface Photoiniferter Graft Polymerization**	70
2.1	Principle	70
2.2	Experimental Evidences of Livingness	74
2.3	Graft Thickness–Gradient Surface	77
2.4	Block-Graft-Copolymerized Surface	78
2.5	Regiospecific Graft-Polymerized Surface	79
3	**Highly Spatioresolved Graft-Chain Architectures: Analogue Models of Tree Growth Progression**	80
3.1	Model I: Hyperbranch Architecture	82
3.1.1	Variable Daughter Chain Lengths but Fixed Parent Chain Length and Degree of Branching	82
3.1.2	Variable Parent Chain Lengths and Variable Daughter Chain Lengths	83
3.1.3	Variable Degrees of Branching but Fixed Parent and Daughter Chain Lengths	83
3.2	Model II: Multigeneration Hyperbranched Graft Architecture	84
3.3	Model III: More Complexly Shaped Hyperbranch Architectures	87
3.4	Model IV: Terminal Endcapping with Functional Group	87
4	**Physicochemical Aspect of Swollen Graft Layer**	90
5	**Biomimetic Surface Architecture**	93
5.1	PC-Group-Endcapped Graft Architecture	93
5.2	Albuminated Graft Architecture	94
6	**Surface Derivatization via Cross-Recombination**	95
7	**Biomedical Applications**	98
7.1	Micropatterned Tissue Formation	98
7.2	Multimicroprocessed Surfaces for Cell Adhesion and Proliferation	98
7.3	High-Throughput Screening for Tissue Compatibility	101
7.3.1	Macrophage and its Fused Foreign-Body Giant Cell	103
7.3.2	Cytokine Production	104
8	**Conclusions**	104
	References	105

Abstract The photoiniferter polymerization method proposed by Otsu et al. was utilized to generate well-controlled graft polymer chains on a surface. The "livingness" of graft chains, coupled with the inherent nature of photochemical processing, enables the development of complex graft-polymerized surface designs with controlled graft-chain length and composition, regiospecific addressability and high-dimensional precision. As an extension of the advantageous features of the "quasi-living" nature of polymerization, precise control technology for surface graft-chain architectures, which show multibranching, a fractal hierarchy and a gradient segmental density, was elaborated. The logical programmed morphogenesis approach was discussed, and a high degree of graft-chain architectures was demonstrated as if these resemble the spatiogeometric analogue models of growing trees with diverse morphologies. The confocal laser scanning microscopic measurement for dye-stained grafted surfaces and the force–distance curves of atomic force microscopy provided some physicochemical and structural insights into graft architectures. Under appropriate conditions, the cross-recommendation reaction of two different dithiocarbamate derivatives enabled the development of a novel surface derivatization method. Microprocessed surfaces with multigraft polymers in different regions and with different chain lengths enabled differentiation of regiospecific cell adhesion and proliferation potentials and cellular functions in one sample, which provides high-throughput screening for the biocompatibility of designed medical devices.

Keywords Biomedical application · Iniferter · Living polymerization · Surface graft architecture

1
Introduction

Recent advanced therapeutic procedures using biomaterial-based medical devices require the absence of adverse reactions derived from foreign-material-induced body defense mechanisms [1]. The fate of medical implant devices in contact with blood or tissues is primarily determined by the nature of a surface, including its chemical composition, structure and properties. For example, whether an implanted artificial organ is accepted or rejected by the body largely depends on its surface design. Surface graft architectures on blood- and tissue-contacting surfaces, when properly designed, minimize or reduce the degree of activation of the body's defense mechanisms. In particular, it has been demonstrated that proton-accepting, nonionic hydrophilic graft layers, such as poly(ethylene glycol) or poly(acrylamide) suppress protein adsorption and cell adhesion at least for a short period of contact with blood or tissue fluid as implanted [1]. The proposed mechanisms underlying such phenomena have been discussed, from the thermodynamical adhesion aspect including the free-energy change and entropy considerations upon adsorption, the water structuring state of hydration/dehydration, the mechanistic aspect of thermally induced chain mobility and the interfacial structural aspect [2]. Alternative approaches include the immobilization of bioinert biomacromolecules (covalent fixation of bioinert proteins such as albumin) [3] or graft polymers containing a phosphatidylcholine (PC) group [4], by which a biomimicking cellular membrane surface is created.

Surface modification can be achieved by the surface derivatization of functional-group-bearing low-molecular-weight substances, the grafting of polymer to and from the surface and amphiphilic polymer coating. The main theme of this article is focused on initiator-transfer-terminator (iniferter)-based "grafting-from-surface" and "derivatization-on-surface" approaches aiming at precision surface architectures, which are primary determinants of the biocompatibility of medical devices.

Conventional graft polymerization techniques followed by plasma, corona discharge or ozone oxidation treatments have been developed over the years [5]. However, the use of these techniques, which allow radical polymerization induced by the thermal decomposition of free radical precursors immobilized or formed on surface regions, cannot provide precision graft architectures due to the extremely high reactivity of free radicals. Therefore, the regiospecificity of graft-polymerized surfaces and precision graft-chain architectures, including their surface graft-chain density, spatiosegmental density, segmental or graded block chains, molecular shape or configuration including the hyperbranching and derivatization of the functional terminal groups of graft chains, cannot be controlled or realized in principle.

To tackle these problems, iniferter polymerization, which was pioneered by Otsu and coworkers in the early 1980s, was extensively studied to develop a new method of radical polymerization aimed at controlled polymerization derived from "living" nature [6–8]. Based on numerous studies by Otsu's group, which dealt with solution polymerization, recent studies have focused on surface graft architectures [9–33]. Using the characteristic features of iniferter-based photopolymerization, graft-chain architectures can be easily manipulated in the growing, branching and endcapping stages, as will be described later (Scheme 1).

This article is intended to summarize new strategic and technological aspects of photoiniferter-based graft architectures in conjunction with biomedical applications as follows.

1. First, the general features of surface photoiniferter graft polymerization initiated by an iniferter immobilized or derivatized on a surface is described. Under appropriate conditions, this chemistry allows the realization of the "quasi-livingness" of graft polymerization and the regiospecific addressability of grafted regions and controlled graft-chain length.
2. Second, a new concept of precise and deliberate control of the macromolecular geometry of graft chains (spatiogeometry of graft architecture) is presented, which may be considered as the spatiogeometric analogue model of tree growth or the fractal configuration model. Precision graft architectures are demonstrated.

3. Third, a newly developed iniferter-based surface derivatization based on cross-recombination reaction is demonstrated.
4. Finally, biomedical applications aiming at controlled protein adsorption and cell adhesion on iniferter-driven surface graft architectures, by which a high-throughput screening of biocompatibility can be materialized, are presented.

2
Surface Photoiniferter Graft Polymerization

2.1
Principle

Iniferter polymerization using dithiocarbamate photolysis chemistry generates a radical pair: one radical (alkyl radical) is capable of initiating vinyl polymerization, and the other (dithiocarbamyl radical) rarely initiates such polymerization (Scheme 1(1)). Because of the inherent extremely high reactivity between these radicals, spontaneous recombination occurs to produce a "dormant" species (dormant termination); however, upon photolysis, a pair of active radical pair species are again regenerated. This enables the realization of the quasi-livingness by which polymerization repeatedly reinitiates during photoirradiation. The impairment of livingness occurs during recombination between two alkyl radicals (bimolecular termination), which is enhanced at a high iniferter concentration. Thus, the sequential reactions of (1) photocleavage to generate a radical pair, (2) monomer addition and (3) recombination proceed with time.

Such "ideal" polymerization behavior is realized only under the condition in which a counter-dithiocarbamyl radical exists at the active propagating end. If an active propagating end loses a counter-radical, this radical becomes a free radical, resulting in the complete loss of the livingness, similarly to conventional radical polymerization. Such an ideal situation, which usually occurs at an early stage of polymerization (relatively low conversion), is influenced by the type of solvent used (for example, the degree of solvation at an active radical pair and the ease of chain transfer to the solvent). To suppress the diffusion of dithiocarbamyl radicals from the active propagating end, the addition of N,N,N',N'-tetraethylthiuram disulfide, which enhances the concentration of counter-dithiocarbamyl radicals upon its photolysis [6–8], is effective. Considering the kinetic aspects of the elemental reaction steps involved in iniferter-based polymerization, diffusion-controlled propagation suppresses the livingness of a system, whereas diffusion-controlled bimolecular termination enhances such livingness. The livingness of a system is also enhanced by diffusion-controlled dormant termination. When diffusion-

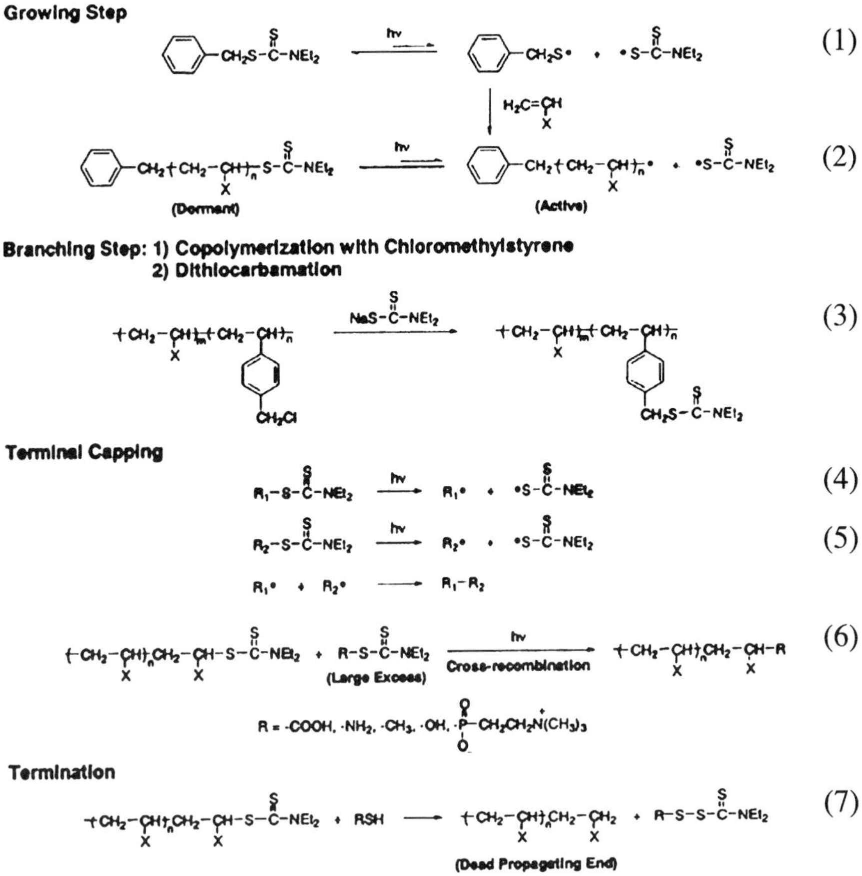

Growing Step

$$\text{PhCH}_2\text{S-C(=S)-NEt}_2 \xrightarrow{h\nu} \text{PhCH}_2\cdot + \cdot\text{S-C(=S)-NEt}_2 \quad (1)$$

$$\downarrow H_2C=CHX$$

$$\text{Ph-CH}_2\text{-[CH}_2\text{-CHX]}_n\text{-S-C(=S)-NEt}_2 \xrightarrow{h\nu} \text{Ph-CH}_2\text{-[CH}_2\text{-CHX]}_n\cdot + \cdot\text{S-C(=S)-NEt}_2 \quad (2)$$

(Dormant) → (Active)

Branching Step: 1) Copolymerization with Chloromethylstyrene
2) Dithiocarbamation

(equation 3: copolymerization of CH₂=CHX with chloromethylstyrene followed by dithiocarbamation with NaS-C(=S)-NEt₂ to give pendant CH₂-S-C(=S)-NEt₂ groups) (3)

Terminal Capping

$$R_1\text{-S-C(=S)-NEt}_2 \xrightarrow{h\nu} R_1\cdot + \cdot\text{S-C(=S)-NEt}_2 \quad (4)$$

$$R_2\text{-S-C(=S)-NEt}_2 \xrightarrow{h\nu} R_2\cdot + \cdot\text{S-C(=S)-NEt}_2 \quad (5)$$

$$R_1\cdot + R_2\cdot \longrightarrow R_1\text{-}R_2$$

$$\text{-[CH}_2\text{-CHX]}_n\text{CH}_2\text{-CH-S-C(=S)-NEt}_2 + R\text{-S-C(=S)-NEt}_2 \text{ (Large Excess)} \xrightarrow[\text{Cross-recombination}]{h\nu} \text{-[CH}_2\text{-CHX]}_n\text{CH}_2\text{-CH-R} \quad (6)$$

$$R = \text{-COOH, -NH}_2, \text{-CH}_3, \text{-OH, -P(=O)(O}^-\text{)CH}_2\text{CH}_2\text{N}^+(\text{CH}_3)_3$$

Termination

$$\text{-[CH}_2\text{-CHX]}_n\text{CH}_2\text{-CH-S-C(=S)-NEt}_2 + \text{RSH} \longrightarrow \text{-[CH}_2\text{-CHX]}_n\text{CH}_2\text{-CH}_2 + R\text{-S-S-C(=S)-NEt}_2 \quad (7)$$

(Dead Propagating End)

Scheme 1 Elemental reactions of DC chemistries used for design of controlled graft architecture

controlled bimolecular termination is negligibly small, the overall effect of diffusion-controlled phenomena in iniferter polymerization is to enhance the livingness of a system. In general, iniferter-based polymerization appears to proceed not via chain reactions but rather via successive reactions in which a chain slowly grows with photoirradiation time.

Assuming perfect living polymerization in which all polymer chains are induced to initiate the simultaneous propagation of chains without any termination and chain transfer reactions, in such idealized scenarios, the number-average molecular weight, M_n, can be estimated using

$$M_n = C \times [M_0] \times [I]^{-1} \times M_{\text{monomer}}, \quad (1)$$

where C is the conversion coefficient, M_0 is the initial number of moles of the monomer, I is the number of moles of the initiator and M_{monomer} is the molecular weight of the monomer. The conversion coefficient is proportional to iniferter concentration and photoirradiation time (t), expressed as

$$C \propto [I] \times t. \tag{2}$$

Therefore, livingness is validated by analyzing the linear conversion–time, conversion–molecular weight and conversion–iniferter concentration relationships. However, such an interpretation appears to be too simple to describe the whole process of iniferter-based radical polymerization, which is far more complex than expected.

More discussion on the relative reactivities on bond dissociation of initiation and propagation steps on the molecular level is needed for the complete understanding of this quasi-living polymerization, which has been recently reported by Ishizu et al. [9–11]. Livingness in this photopolymerization depends on the photodissociation ability of the C – S bond of dithiocarbamate (note that S – C(= S) bond dissociation leads to nonliving, typical free-radical polymerization). From the results of density function theory calculations for several model compounds, Ishizu et al. [9] concluded that steric factors are important in determining bond dissociation energy. Figure 1 shows the C – S bond dissociation energies and bond lengths of several model compounds. That is, C – S bond dissociation energy increases in the order of tertiary<secondary<primary carbon, and an electron-withdrawing group (e.g., nitrile or carbonyl) attached to a carbon reduces bond dissociation energy. Ishizu et al. [10] anticipated that this should be reflected in larger bond lengths of the ground state in which steric factors have greater effects. The calculations predict that C – S bond dissociation ability reduces in the order of the compounds $5 > 4 \geq 3 > 2 > 1$ (Fig. 1).

According to their recent reports [11], **5** provides livingness of polymerization for styrene (ST) and methyl methacrylic acid (MA). Irrespective of the type of monomer, the initiator efficiency f was over 0.9, and the polydispersity index, Mw/Mn (Mw: weight-average molecular weight, Mn: number-average molecular weight), was close to unity (approximately 1.2). The polymerization rate was very low (on the order of an hour), i.e., a very slow

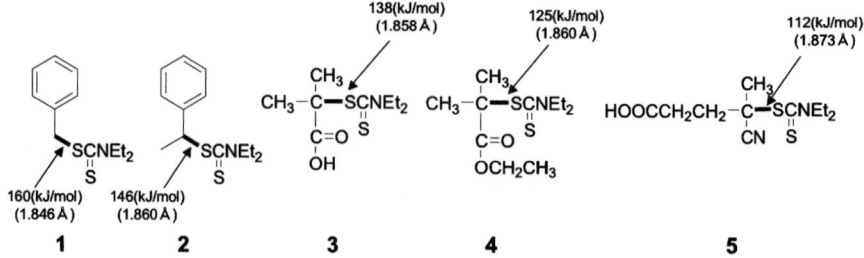

Fig. 1 The C – S bond dissociation energies and bond lengths for several model compounds, calculated assuming a hemolytic bond cleave

monomer addition reaction occurs. These polymerization characteristics of high initiation efficiency and slow polymerization rate observed for the polymerization of ST and MA using 5 may be explained as follows: the first photolysis prior to monomer addition proceeds at a very high rate due to a very low bond dissociation energy (112 kJ/mol; see Fig. 1) and the polymerization rate, determined by the bond dissociation energy of the ST- or MA-added iniferter (RMn-DC: Scheme 2) is low due to the relatively high bond dissociation energy, which is deduced from the structural analogies of the iniferters 2 and 3, as shown Fig. 1.

On the other hand, 3 provides a nonliving radical polymerization behavior for ST: its f is 0.32 and its Mw/Mn is approximately 2.0, regardless of its molecular weight, whereas it induced the living radical polymerization of 2-hydroxyethyl methacrylate (HEMA), N-isopropylacrylamide (NIPAM) and methyl methacrylate (MMA) at high polymerization rates (several tens of minutes for complete polymerization). This may be interpreted as follows: since bond dissociation energy directly determines the ease of photolysis, an iniferter with a C–S bond with a lower dissociation energy induces a faster initiation reaction. For a macroiniferter, in which a monomer was added to the iniferter (the analogies found for 2 for ST and 4 for MMA], a poly (MMA)-iniferter has a much higher polymerization rate than poly(ST)-iniferter.

Based on the above results, an ideal "living radical polymerization" may proceed under appropriate reaction conditions as follows (Scheme 2).

1. Initiation step: the decomposition rate constant (kd) of an iniferter ($R - DC$) is much larger than that (kd') of a monomer-added iniferter ($RM_n - DC$). This enables the maximization of iniferter efficiency.
2. Propagation step: the following equation holds under the steady state. Monomer consumption rate can be written as

$$\frac{d[M]}{dt} = kp[RM_n\cdot][M] \tag{3}$$

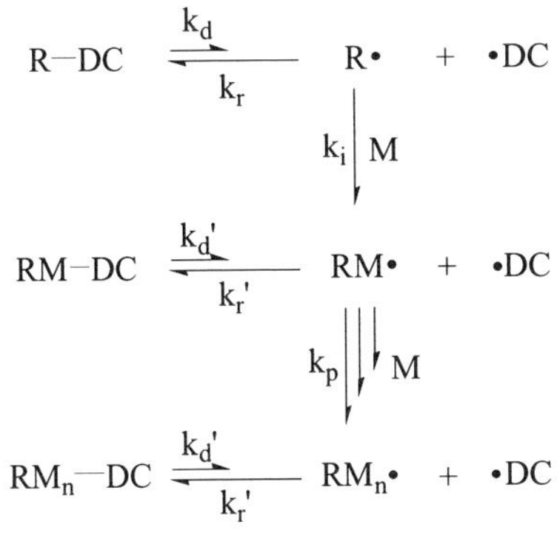

$$R-DC \underset{k_r}{\overset{k_d}{\rightleftarrows}} R\cdot + \cdot DC$$
$$\downarrow k_i \, M$$
$$RM-DC \underset{k_r'}{\overset{k_d'}{\rightleftarrows}} RM\cdot + \cdot DC$$
$$\Updownarrow k_p \, M$$
$$RM_n-DC \underset{k_r'}{\overset{k_d'}{\rightleftarrows}} RM_n\cdot + \cdot DC$$

Scheme 2 Kinetic model of photolysis, recombination and polymerization

where kp denotes the monomer addition rate constant.

Since $kd'[RMn - DC] = kr'[RM_n\cdot][\cdot DC]$ and $[\cdot DC] = [RM_n\cdot]$ (where kr' is the recombination rate constant), Eq. 3 can be rewritten as

$$\frac{d[M]}{dt} \alpha kp \left[\frac{kd'}{kr'}\right]^{1/2} [M] . \tag{4}$$

Therefore, monomer consumption rate largely depends on kd': that is, a larger kd' leads to a higher polymerization rate. If kd' is much larger than kd, the predicted polymerization behavior is characterized by a low initiator efficiency and a wider polydispersity.

2.2
Experimental Evidences of Livingness

Surface photograft polymerization was conducted under ultraviolet (UV) irradiation onto iniferter-derivatized surfaces or iniferter-containing co-polymer-coated surfaces, as shown in Scheme 3. Many studies have demonstrated the livingness of graft polymerization under appropriate reaction conditions, that is, the linear correlation of the molecular weight of the grafted chain with photoirradiation time as well as conversion coefficient [9–20]. The in situ real-time continuous monitoring of livingness was achieved using a quartz crystal microbalance (QCM), which allows real-time detection of the increase in the weight of a QCM electrode on the nanogram order as a change in resonance frequency, which is directly converted to the weight of polymerized graft chains. Figure 2a shows an example of ST graft polymerization on a QCM electrode [20]. The amount of grafted polymers linearly increased with UV irradiation time. The apparent polymerization rate lin-

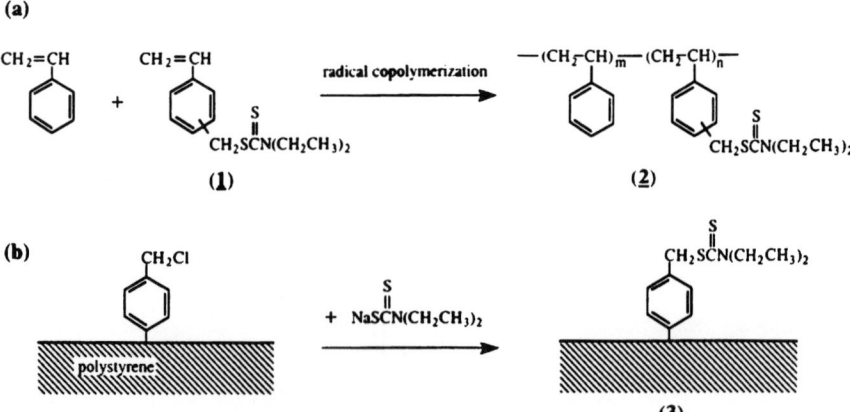

Scheme 3 a Iniferter-derivatized polymer-coated surface. b Iniferter-derivatized surface

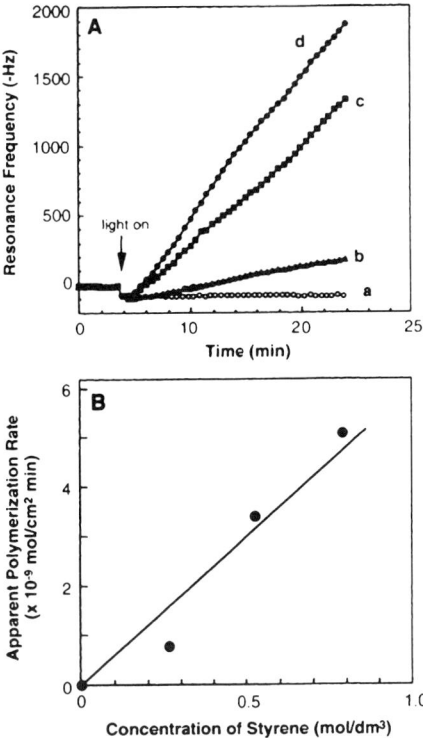

Fig. 2 **A** Oscillator frequency changes with graft copolymerization of styrene (ST) on the dithiocarbamated copolymer-coated quartz crystal microbalance (QCM) in methanolic solution with UV irradiation (light intensity: 5 mW cm^{-2}). Concentration of ST: *a* 0, *b* 0.26, *c* 0.5, *d* 0.76 mol dm^{-3}. **B** Relationship of oscillator frequency change calculated from **A** with concentration of ST

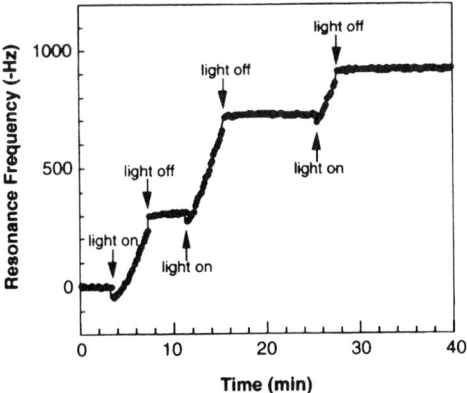

Fig. 3 Oscillator frequency change of the dithiocarbamated copolymer-coated QCM in a methanolic solution of ST (concentration of monomer: 0.5 moldm^{-3}) with repeated UV irradiation (light intensity: 5 mWcm^{-2}) and cessation

early increased with ST concentration (Fig. 2b), which is in accordance with Eq. 4. Figure 3 shows the change in resonance frequency with or without UV irradiation. UV irradiation resulted in an immediate monotonic decrease in resonance frequency, but on cessation of irradiation, a slight change in resonance frequency was observed. Repeated cycles of irradiation and cessation produced very similar changes in resonance frequency without appreciable fatigue. Such on–off behavior is evidence that the dormant species of growing chain ends are activated upon UV irradiation without significant loss of livingness [20]. The linear relationships of weight gain with irradiation time and photointensity as well as monomer concentration proved that surface graft polymerization proceeds via the equilibrium reaction between the dormant and active species of propagating chain ends, as schematically shown in Scheme 2.

Topological images of growing graft layers were recorded by scanning the surface with the sharp tip of an atomic force microscope (AFM) [22]. Fig-

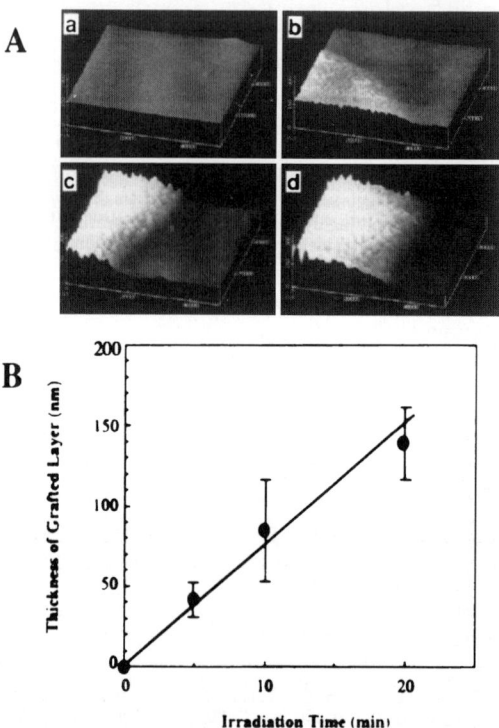

Fig. 4 A Atomic force microscope (AFM) images, projected at 30° for viewing, of nontreated dithiocarbamated polymer film surface (*a*) and the PST(polystyrene)-grafted surface by UV irradiation through the lattice-patterned projection mask for 5 (*b*), 10 (*c*), and 20 min (*d*). **B** Relationship between the average thickness of the PST-grafted layers and UV irradiation time

ure 4a shows the AFM images, including a top view and a cross-sectional view, at the edge of the photoirradiated and nonirradiated regions. The growing polymerized ST layer was observed with photoirradiation time. The average thickness of the graft layer, determined from the cross-sectional images, increased almost linearly with photoirradiation time (Fig. 4b), suggesting that the graft chain propagates steadily with time. After 20 min of UV irradiation, the measured thickness in this particular case was approximately 140 nm.

On the basis of the above lines of evidence, coupled with the inherent nature of photochemical processing, three graft-polymerized surfaces are described below: a graft thickness-gradient graft-polymer, a block-graft-polymerized surface, and a regiospecific and dimensionally precise graft-polymer on surfaces.

2.3
Graft Thickness–Gradient Surface

In addition to the above real-time weighing of growing grafted chains, an alternative method of demonstrating the growth of graft chains is the determination of gradient chain-length, which is realized by changing either photoirradiation time or photointensity. The former was achieved by photoirradiation using a moving shutter, and the latter by photoirradiation using a gradient neutral-density filter [17]. The following is an example of the latter case. A graft thickness-gradient surface in which graft yield or possibly graft-chain length changed unidirectionally was prepared using a stripe-patterned

Fig. 5 Fluorescence micrograph of surface photo-graft-copolymerized with (N-dimethylamino) propyl acrylamide methiodide (DMAPAAmMeI) by UV irradiation through the stripe-patterned projection mask and the neutral-density filter and subsequently stained with rose bengal, and the three-dimensional image, *b* of the distribution of the florescence intensity in the area shown in *a*

projection mask coupled with a gradient neutral-density filter, which is a cut filter with a unidirectional continuous change in light intensity. A cationic monomer was polymerized on an iniferter-derivatized surface through the gradient filter, and then stained with an anionic dye (rose bengal). Using confocal scanning laser microscopy (CLSM), a fluorescence micrograph of the treated surface (Fig. 5a) was obtained. The distribution of the fluorescence intensity derived from rose bengal in the square area in Fig. 5b was superposed on the micrograph, and the fluorescent intensity gradients of the dye-stained graft chains are clearly observed. This is in good agreement with the gradient light transmittance of the filter used and indicates that graft-chain length gradually varies with the distance from the nonirradiated end of the film.

2.4
Block-Graft-Copolymerized Surface

The graft block chain can be obtained by a sequential procedure: (1) photoirradiation in the presence of the first monomer A, (2) exchange with the second monomer B and (3) photoirradiation. This leads to the formation of an AB-type diblock graft chain. The repeated cycles of procedure or the use of the third monomer C produces AB multiblock or ABC-type triblock graft chains, respectively. Figure 6 shows the cross-sectional transmission electron micrograph (TEM) image of the block-graft-copolymer formed on the surface, composed of two polymers, poly(N,N)-dimethylaminoethyl acrylamide (DMAEMA) and PST, proving that well-phase- separated block-graft layers are produced by two-step photoiniferter polymerization. Such a block-graft copolymer formed on a surface was demonstrated for a functional biocompatible and drug-eluting surface [22].

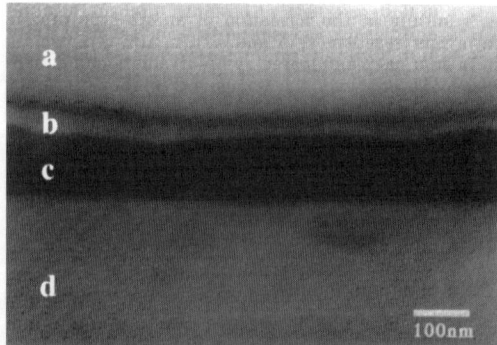

Fig. 6 Transmission electron micrograph (TEM) of the cross-section of a PDMAAm(polydimethylacrylamide)-b-PST(polystyrene) block-graft-copolymerized surface stained in iodine vapor. *a* Spurr's resin, *b* PST layer, *c* PDMAm layer, *d* dithiocarbamate (DC)-derivatized PST film

2.5
Regiospecific Graft-Polymerized Surface

When UV light is irradiated through a photomask, photoiniferter polymerization occurs only at photoirradiated areas, resulting in the formation of regionally dimension-controlled surfaces. An example is shown in Fig. 7a. Using a stripe-patterned projection mask and two monomers, a crosshatched micropatterned surface was formed. That is, after an initial stripe grafted region was formed, another stripe grafted pattern perpendicular to and overlaid on the first grafted surface was formed using the mask placed on the surface after rotating it 90° relative to its orientation in the former case. The crosshatched graft surface regions, which are the intersection of the two grafted layers, consist of block graft copolymers. The surface thus prepared was composed of four types of region: nongrafted, poly(DMAEMA)-grafted, poly(MA)-grafted, and poly(DMAEMA-b-MA)-grafted (intersec-

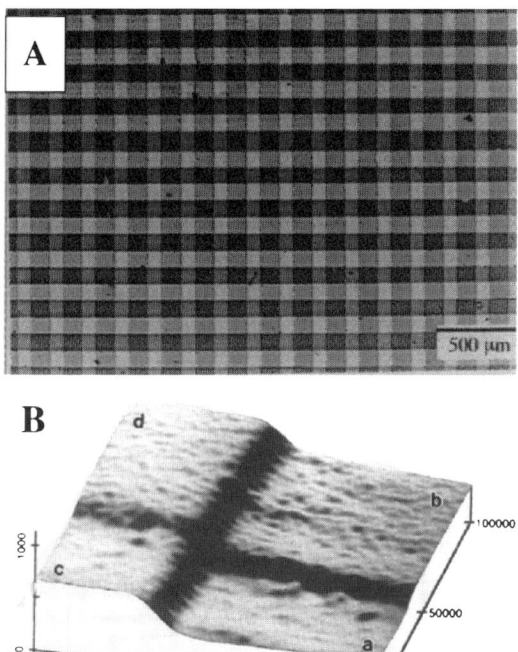

Fig. 7 **a** Micrograph of a cross-lattice patterned surface upon photo-graft copolymerization using MA or DMAPAAm by UV irradiation through the stripe-patterned projection mask. **b** AFM image of cross-hatched graft surface: *a* nontreated dithiocarbamated PST film, *b* poly(DMAPAAmMeI-b-MA)grafted region, *c* poly(MA)-grafted region, *d* poly(DMAPAAmMeI-b-MA)grafted region

tions) regions. A topographic image of the stripe intersection portions prepared was obtained by AFM (Fig. 7b), indicating that the micrometer-ordered regiospecificity and dimensional precision of the grafted region is assured [15–17].

This regiospecificity was utilized for the preparation of microarrays in which different types of graft chain are grafted in respective arrays. This was utilized as high-throughput cell assay substrates for screening the adherence of macrophages and its transformation to giant cells [31–33]. Moreover, microarrays consisting of a gradient graft chain in each array enabled us to simultaneously define the cell adhesion potential dependences on physicochemical properties and compositions, and graft-chain length [22]. These will be described in more detail later. Recently, a new surface-engineering technology utilizing a combination of photoiniferter and photolithographic techniques has been developed by Peppas et al. [15], which enables the preparation of unique micropatterned surfaces with high aspect ratios. Such surfaces may be used for cellomics and biosensors.

3
Highly Spatioresolved Graft-Chain Architectures: Analogue Models of Tree Growth Progression

Tree growth progresses with trunk or stem length and the degree of branching from boughs to twigs to leaves results in the formation of dendrite-like fractal patterns or dendritic topologies. If surface graft polymerization proceeds in a tree architecture manner (morphogenesis), a logical synthetic approach toward the divergent, generational, dendritic growth of graft architectures may be feasible. The key is to utilize the photochemistry of dithiocarbamate as follows. The elemental reactions enabling growth, blocking, branching, termination and terminal capping are represented in Scheme 1.

1. Growth control: the growth stage is controlled by adjusting photoirradiation intensity and photoirradiation time as well as monomer concentration (Fig. 8a). The control of graft-chain length was already verified as described above.
2. Branching control: branching can be realized using chloromethylstyrene (CMS) as a comonomer [Scheme 1,(3) and (4)]. After copolymerization (formation of stem or parent chain), the CMS unit is converted to a dithiocarbamate (DC) group using sodium N,N-diethyldithiocarbamate at room temperature. The subsequent photoirradiation in the presence of a monomer allows the formation of a branch (or daughter) chain initiated from the DC-derivatized site. A granddaughter chain can also be incorporated by sequential copolymerization with CMS during the preparation of the daughter chain and the subsequent dithiocarbamation, followed by

photopolymerization. Hyperbranching is realized by repeated cycles of the sequential procedure (Scheme 4 and Fig. 8(b)).
3. Growth termination and endcap modification: the complete termination of a growing chain end is carried out by either of the two reactions in Scheme 1, (6) and (7). Photoirradiation in the absence of a monomer but

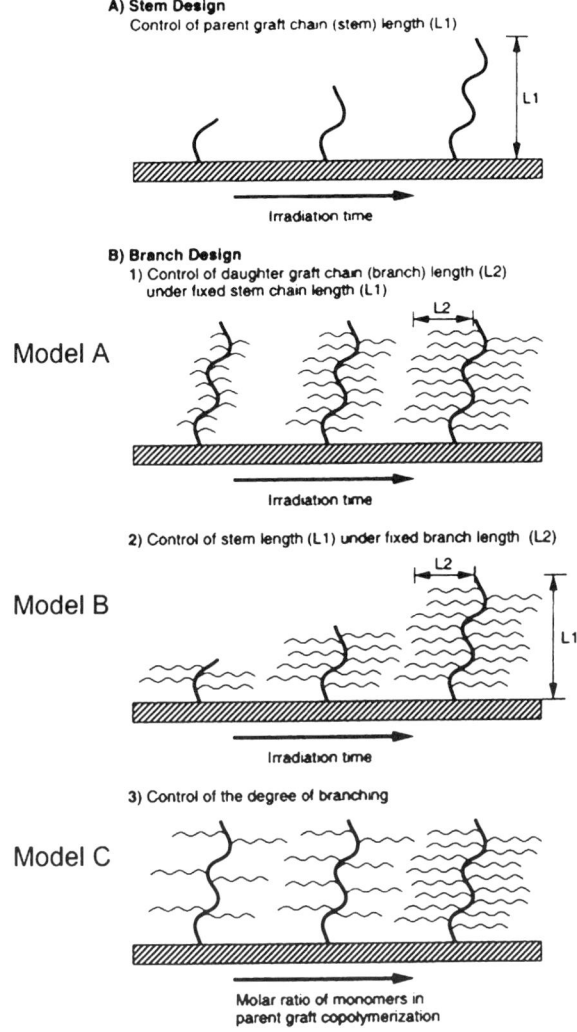

Fig. 8 Schematic representation of the synthetic approaches of stem and branch designs. *Model A* is prepared under a fixed parent-chain length but different daughter-chain lengths. *Model B* is prepared under a fixed daughter-chain length but different parent-chain lengths. *Model C* is prepared under fixed parent-chain and daughter-chain length but different daughter-chain densities

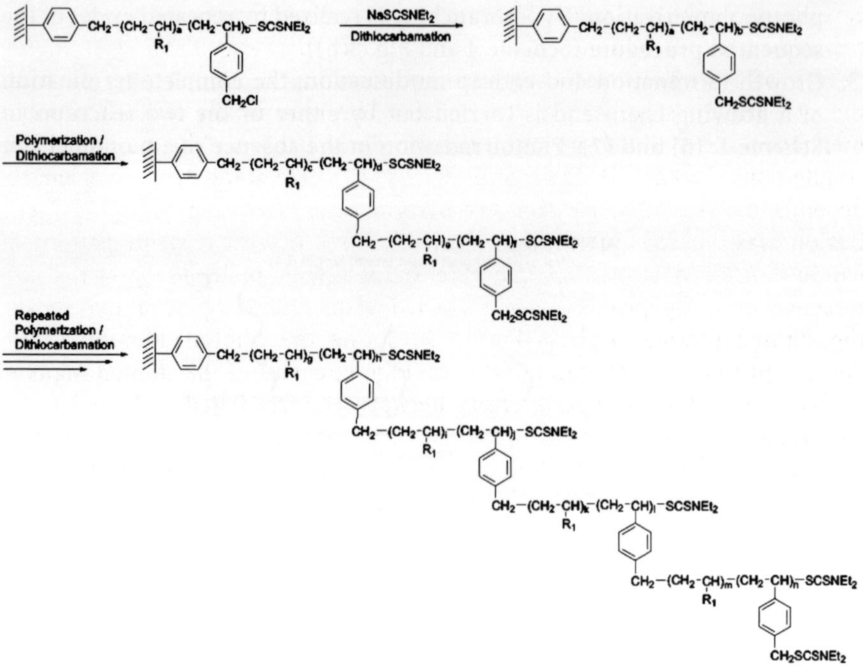

Scheme 4 Hyperbranched multigeneration graft architecture based on sequential repeating process of copolymerization with chloromethyl styrene (CMS) as a comonomer and subsequent dithiocarbamation

in the presence of alkanethiol results in the transformation of a dormant species to a "dead" species at the growing chain end. On the other hand, our recent study has revealed that, in the presence of a large excess of alkyl dithiocarbamate, photoirradiation results in the cross-recombination between a radical at a propagating chain end and an alkyl radical derived from an externally added dithiocarbamate, resulting in the formation of a dead species (cross-recombination). When an alkyl group is properly selected, a desired functional group can be incorporated into the terminal end of each branch (Scheme 1: terminal capping).

3.1
Model I: Hyperbranch Architecture

3.1.1
Variable Daughter Chain Lengths but Fixed Parent Chain Length and Degree of Branching

CMS was copolymerized with N,N-dimethylacrylamide (DMAm) to form a parent (or stem) chain with a fixed CMS content. Subsequently, dithiocarbamation with sodium *N,N*-diethyldithiocarbamate at room temperature resulted in the formation of a multiply iniferter-derivatized stem chain. The degree of branching was controlled by adjusting monomer feed ratio at the preparation stage of the stem chain. The reinitiation of graft copolymerization with MA from the iniferter derivatized on the parent chain resulted in a branched graft architecture in which daughter chain length depends on second-stage photoirradiation time. Two-step photopolymerization was carried out using a projection mask with a stripe-patterned window. After daughter chains were stained with malachite green, fluorescence intensity profiles were recorded using CSLM by scanning across the stained grafted regions. Figure 9a shows the photoirradiation time-dependent intensity images, and the average intensity of the stained regions (difference in fluorescence intensity between the irradiated and nonirradiated regions) increased proportionally with photoirradiation time, indicating that the chain length of daughter graft chains on the fixed stem design (chain length and degree of branching) increases with photoirradiation time.

3.1.2
Variable Parent Chain Lengths and Variable Daughter Chain Lengths

Photoirradiation time for controlling stem design with CMS was varied but the photoirradiation time for controlling the branch design with MA was fixed. Figure 9b shows the fluorescence intensity profiles of the malachite-green-stained, regionally grafted surface. The average fluorescence intensity exhibited a linear dependence on the photoirradiation time of the stem design, indicating that parent graft-chain length and fixed daughter chain length increases with photoirradiation time (Fig. 9b).

3.1.3
Variable Degrees of Branching but Fixed Parent and Daughter Chain Lengths

Stems with different contents of the CMS unit were first prepared, followed by dithiocarbamation. The branch design with MA was carried out at a fixed photoirradiation time. The fluorescence intensity profiles indicate that the higher the CMS content in the stem chain, the higher the fluorescence intensity is Fig. 9c. The plot of average fluorescence intensity with CMS content gave a fairly good linear relationship, indicating that fluorescence intensity is proportional to the degree of branching. This linear relationship obtained indicates that the design strategy developed for fine structured graft architectures does operate well.

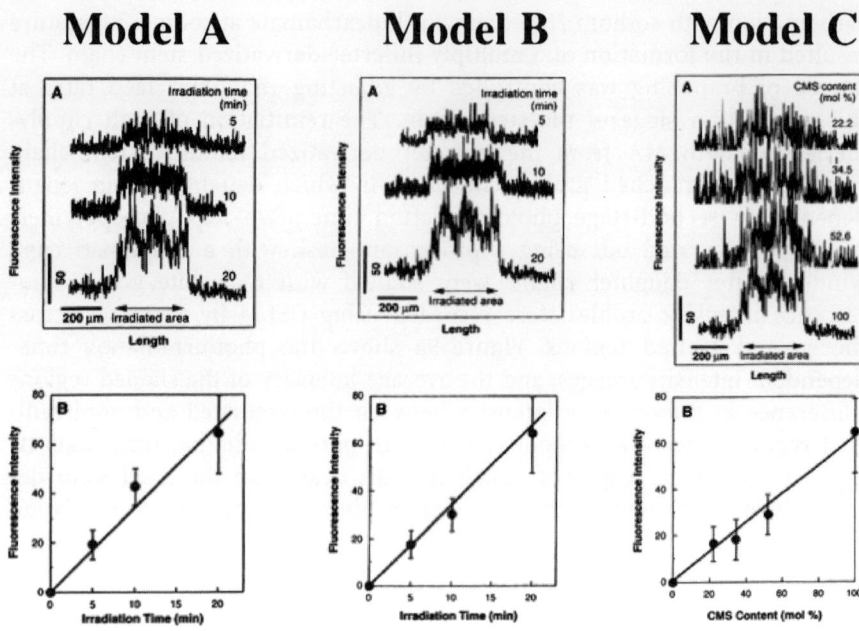

Fig. 9 *Top:* Rose-bengal-stained fluorescence images of photograft-polymerized regions of *Models A, B* and *C* as a function of photopolymerization time or copolymer composition of CMS (see Fig. 8 legend). These indicate that stem-chain length (Model A), daughter-chain length (Model B) and daughter-chain density (Model C) are well-controlled. *Bottom:* Change in fluorescence intensity as a function of photopolymerization time or copolymer composition of CMS *CMS* Chloromethyl styrene

3.2
Model II: Multigeneration Hyperbranched Graft Architecture

A more complex but controlled hyperbranched graft architecture can be prepared by repeated cycles of sequential photo-copolymerization with CMS and dithiocarbamation. A three-generation branched-graft architecture consisting of parent (stem), daughter (branch) and granddaughter (twig) chains was demonstrated (Fig. 10). Regionally different hypergeneration-graft architectures with DMAEMA were prepared stepwise on the dithiocarbamated surface. After its quarterization with methyl iodide, the surface was stained with a dye. Figure 11 shows the micrograph of the stepwisely, regiospecifically prepared surface in which Go is the dithiocarbamated surface (non-grafted), GI is the first-generation (parent or stem) chain, GII is the second-generation (branch or daughter) chain, and GIII is the third-generation (twig or granddaughter) chain. Higher generation graft surfaces were more densely stained, as clearly visualized on the regional multigeneration graft architecture surface.

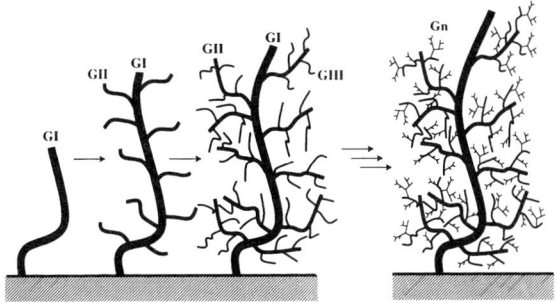

Fig. 10 Schematic drawing of the 1st generation (GI) to *n*th generation (Gn)-graft architectures

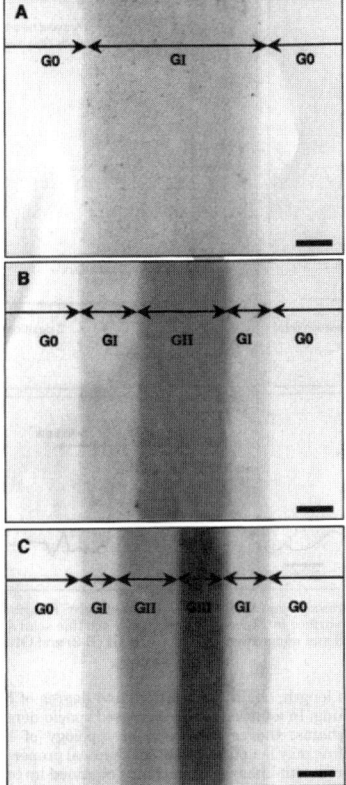

Fig. 11 Visualization of the sequential progress of branching stage by staining with rose bengal. DC-derivatized PST surface (*G0*) was graft-copolymerized with CMS and dimethylaminoethyl acrylamide (DMAEMA), and subsequent quarternization, while narrowing the irradiation area in each polymerization stage (item*, *GI*; item*, *GI+GII*; item*, *GI+GII+GIII*) by the combination of three kinds of masks with linear openings (line widths: 2 mm for GI, 1 mm for GII, 0.5 mm for GIII). *Bar*=0.5 mm

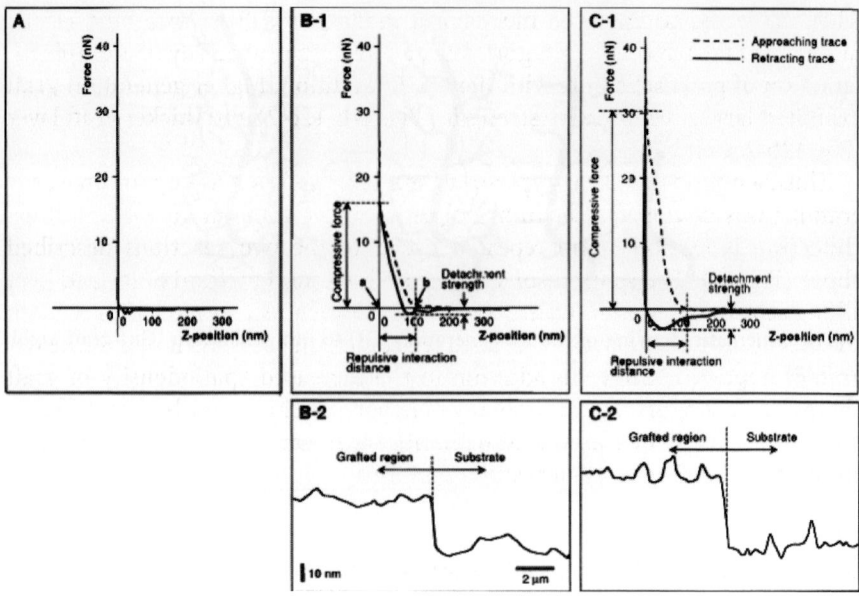

Fig. 12 Force-versus-distance (f-d) curves using atomic force microscope on (A) dithiocarbamated PST surface and CMS-DMAEMA-graft-copolymerized surfaces (B-1; GI surface, C-1; GIII surface in Fig. 11) and their scanned topological features (B-2 and C-2)

On the other hand, structural information on graft chains in water was obtained by analyzing the force-versus-distance (f–d) curve using AFM (Fig. 12). In principle, the f–d curve analysis could provide invaluable, dynamic information on the structure of the swollen graft architecture upon the forced pushing in of a tip into and pulling out from the swollen graft layer, which is described in the detail in the latter part of this review. It was apparent that, as multigeneration progressed, the hyperbranching of the graft architecture was enhanced, resulting in a higher spatiodensity of the graft chain. This may reflect the mechanical properties of swollen graft architecture surfaces. As evidenced by the results of the comparison of the f–d curves for the GI and GIII surfaces, (Fig. 12 a and c-1) with increasing number of generations, a higher steric repulsion was observed as the probe tip came into contact with the outermost graft chain and penetrated into a graft layer, both of which were derived from the high spatiodensity of the graft chain and the configuration or topology of hyperbranching. The degree of hysteresis between approaching and retracting traces, which may be derived from the elastic pushing-out force generated upon compression of graft chains, was larger for the GIII surface than for the GI surface. In addition, the detachment (or adhesive) strength was found to be larger for the GIII surface than for the GI surface. Multiple adhesion jumps were also observed. This suggests that higher generation graft chains combined their interactions with the tip when subjected to enforced mechanical loading, resulting in a high elastic modulus of the water-swollen graft layer and a high probability of multiple interaction of graft segments with the tip. In addition, higher generation graft exhibited larger compressive strength (Fig. 12b-1, c-2) and thicker graft layer (Fig. 12b-2, c-2).

Thus, a multigeneration hyperbranched graft architecture up to three generations was demonstrated. In principle, a higher generation (nth) graft architecture is feasible upon repeated cycles of the two reactions described above (Fig. 10). The spatioresolved graft architectural method described here allows selection from multiple choices of graft architectures in terms of main chain length, average branch–branch length, chain composition, and the degree of hyperbranching. In addition to the increased spatiodensity of graft chains, the configuration or topology of hyperbranching may lead to different mechanical properties compared with those of linear graft chains obtained by conventional radical polymerization methods.

3.3
Model III: More Complexly Shaped Hyperbranch Architectures

Since the propagating chain ends of stem and branch chains in Models I and II have dithiocarbamate groups, both stem and branch chains grow simultaneously, as schematically shown in Fig. 13a. When UV light is irradiated in the presence of alkanethiol but in the absence of a monomer, a propagating dormant species is converted to a dead species due to radical recombination, which cannot be dissociated into active species [Scheme 1 (7)]. Therefore, if the termination step mentioned above is incorporated into repeated cycles of the procedure, a hyperbranching graft architecture with a controlled chain length during the generation of each graft is realized (Fig. 13b). A palmtree-like graft architecture is demonstrated in Fig. 13c, in which CST was copolymerized at the last stage of stem chain preparation, followed by the sequential procedure mentioned above. A fractal topographic graft architecture is shown in Fig. 14, in which topographic evolution up to the fourth generation is schematically shown. In this case, dendritic or cascade-like morphogenesis is achieved through receptive generational steps.

3.4
Model IV: Terminal Endcapping with Functional Group

At the last stage of graft-chain preparation, irrespective of the type of stem or branch, a change in the type of monomer used produces block graft architectures (Fig. 15a). As shown in Scheme 1, if an excess amount of dithiocarbamate with functional groups exists in a monomer-free solution, photoirradiation leads to the cross-recombination between a propagating end of the grafted surface and the alkyl radical generated by the dithiocarbamate exter-

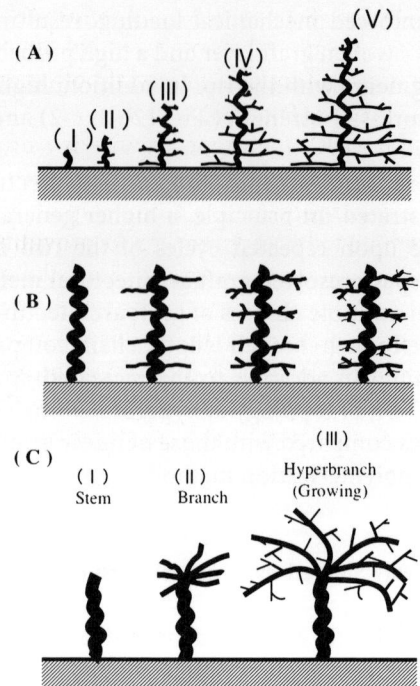

Fig. 13 Hyperbranced graft architecture based on "graft-on-graft" technique. **a** Growing tree model (both stem and branch) **b** Growing hyperbranching model (controlled chain length) **c** Growing hyperbranching at stem end

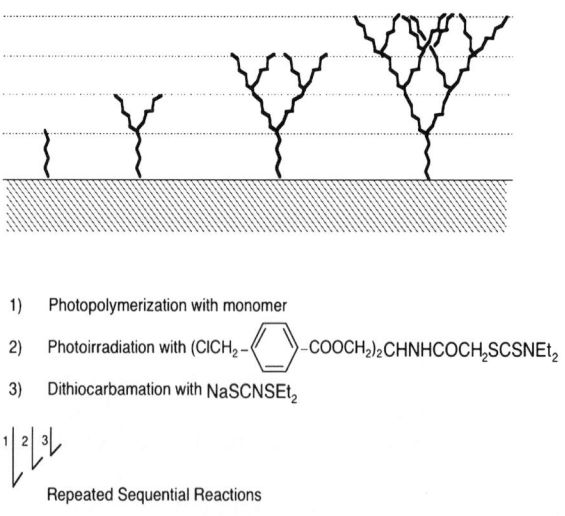

Fig. 14 Growing hyperbranching model (fractal design)

nally added to the solution (Scheme 1(6)). Therefore, if an alkyl radical has functional groups such as amino, hydroxyl, methyl, or carboxyl groups, the terminal ends of the propagating chain ends of both stem and branches are endcapped with these groups (Fig. 15b).

A phosphatidylcholine (PC)-endcapped lipid exists on the outer surface of red blood cells, which do not adhere to any synthetic substrates. Owing to this phenomenon, the PC incorporation on a biocompatible surface design has been extensively attempted. Two approaches to providing PC groups at the terminal ends of each chain were presented. First, the cross-recombination technique was utilized. Using an excess amount of PC-derivatized dithiocarbamate, PC groups were derivatized at the terminal ends after hyperbranched graft chains were prepared, and cross-recombination with dithiocarbamate resulted in terminal endcapping. The other approach is the use of an iniferter with a PC group, which always exist at a propagating chain end [29]. These will be dealt in a later section.

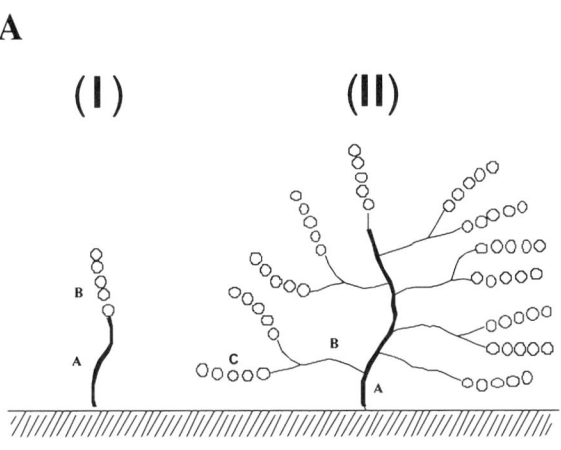

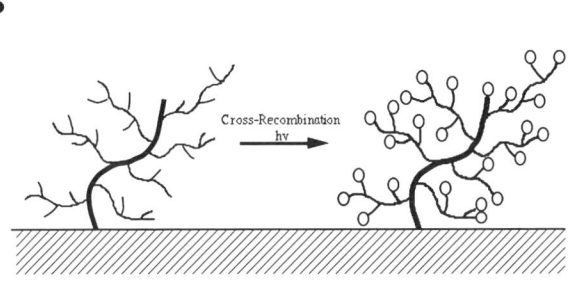

Fig. 15 **a** Hyperbranched block graft architecture. **b** Terminal endcapping via cross-recombination in the presence of DC substance

4
Physicochemical Aspect of Swollen Graft Layer

Recent developments in force measurement techniques using AFM have provided a useful strategy for determining the structural and mechanical features of graft layers in water, as demonstrated in Fig. 12. The AFM f–d curves were conducted for photoiniferter-based photopolymerized poly(DMAm) graft chains in water, which were obtained with varied photoirradiation times (Fig. 16). In the approaching trace of all the f–d curves obtained for the poly(DMAm) surface, repulsive forces derived from the steric interaction of poly(DMAm) were observed [28]. The repulsive interaction distance, defined in Fig. 16, gradually increased with increasing photoirradiation time, i.e., chain length. From the dependence of the repulsive interaction distance on photoirradiation time, it was found that the distance sharply increased in the very short period of photoirradiation, ($\sim$ 10 s) followed by a linear increase after 30–40 s (Fig. 17a). The force at zero position corresponds to

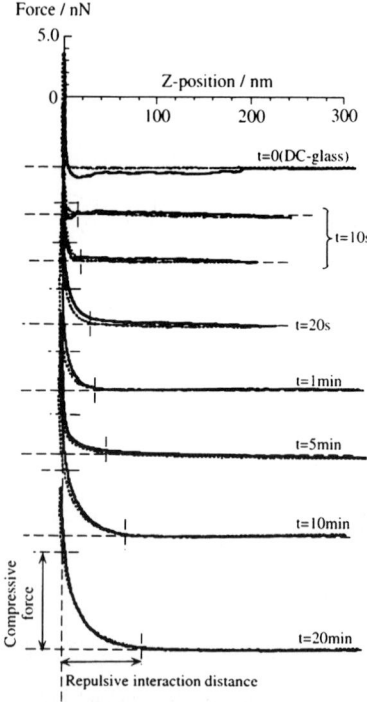

Fig. 16 Representative f-d curves measured for photograft-polymerized surfaces of DMA after photoirradiation for 10 s, 20 s, 1 min, 5 min, 10 min, and 20 min. *Dashed curves* Approaching trace, *solid curves* retracting trace. The repulsion interaction distance increased with photoirradiation time

the full compressive force of the layer, and the repulsion interaction distance represents the degree of compression deformation of the graft layer. On the basis of these definitions, the plot of compressive force against compression deformation (Fig. 17b) shows an approximate linear increase after the deformation, indicating that the elastic behavior of the poly(DMAm) layer approximately conforms to Hooke's law. From these lines of experimental evidence, in the very short period of photoirradiation (10–20 s), a scattered "mushroom" state of graft chains which are scattered on the surface, followed by the monolayer formation of mushrooms (approximately 20 s). The subsequently vertical stretching of graft chains from the surface leads to the formation of a "polymer brush" as schematically shown in Fig. 18. Young's modulus, E, was roughly determined to be approximately 1.2 MPa [31]. This value is considerably larger than those of living cells (0.013–0.5 MPa) and cartilage (0.16–0.6 MPa). Presently, the exact explanation for this large value of Young's modulus of poly(DMAm) brush formed on surface is unknown [28].

On the other hand, poly(N-isopropylacrylamide), poly(NIPAM), has a lower critical solution temperature (LCST) at 32 °C in water; it is water-soluble below LCST but precipitates above LCST. The thermoresponsive

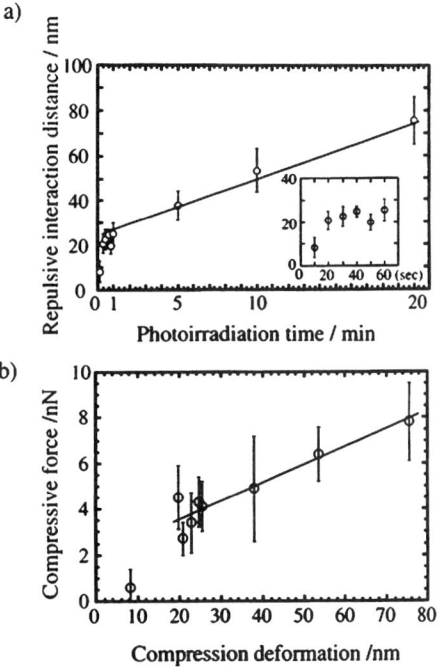

Fig. 17 a Dependence of the repulsive interaction distance between the albumin (*Alb*)-fixed tip and the photograft-polymerized surface of DMAm on the photoirradiation time. A magnification of the plot within 60 s is shown in the *inset*. **b** Dependence of compressive force of the DMAAm graft-polymerized layer on its compression deformation

change in the structure of the poly(NIPAM) graft chain is determined from the f–d curves [29]. Figure 19 shows the approaching trace of f–d curves measured for the AFM tip and poly(NIPAM) grafted surfaces with different chain lengths (graft polymerization under a fixed photoirradiation time but at different monomer concentrations). In all the f–d curves measured at 25 °C, a repulsive force was observed, which increased with increasing monomer concentration, indicating that as graft-chain length increases. The

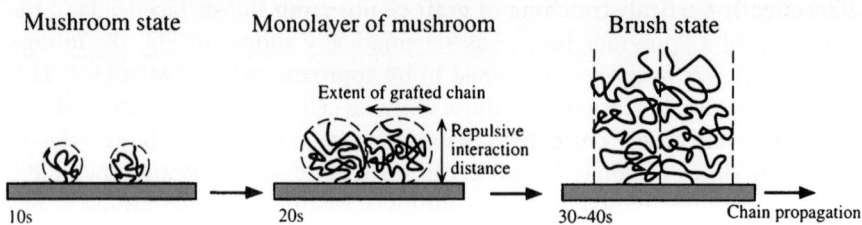

Fig. 18 Schematic of the conformational changes from mushroom to brush state that occur in graft-polymerized PDMAm during the chain propagation process

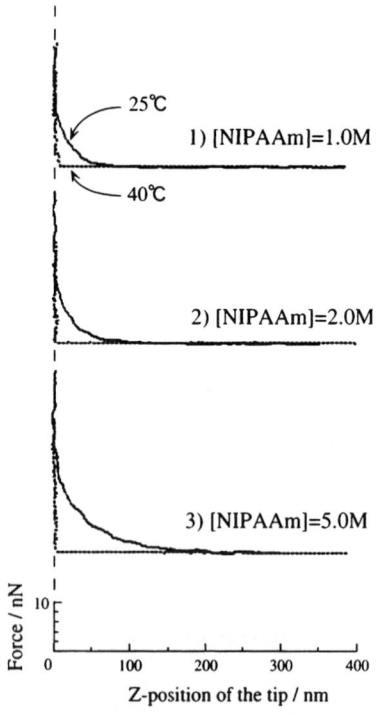

Fig. 19 Representative force-versus-distance curves (approach trace) measured on NIPAAm graft-polymerized surfaces. *Solid curves* 25 °C, *dashed curves* 40 °C

repulsion interaction distance also increases owing to the conformational state of water-swollen graft chains. This is in good agreement with that observed for nonthermoresponsive poly(DMAm), as described above. However, the f–d curves measured at 40 °C exhibited a marked difference: repulsion interaction distance disappeared, irrespective of graft-chain length, indicating the thermoresponsive collapse of the graft chains. Recently, poly(DMAm) graft polymerized from carboxyl group-bearing iniferter immobilized on an AFM tip and substrate surface was prepared, followed by protein immobilization using a water-soluble condensation agent. This allows a precise determination of the protein–protein interaction force [34].

5
Biomimetic Surface Architecture

5.1
PC-Group-Endcapped Graft Architecture

Since the PC group is a polar head of a major lipid component of the cell membrane of red blood cell which is the only nonadherent cell among various blood cell types, a biomimetic approach to incorporating the PC group on the surface at the terminal end of water-soluble graft polymers has been

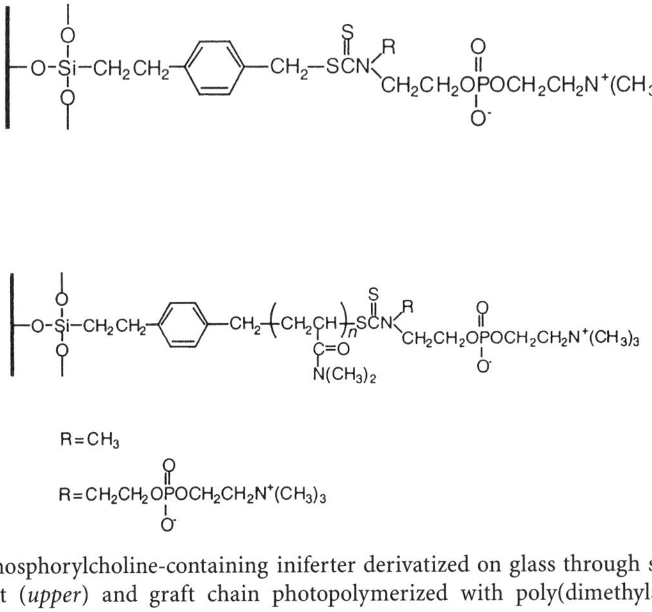

Fig. 20 Phosphorylcholine-containing iniferter derivatized on glass through silane coupling agent (*upper*) and graft chain photopolymerized with poly(dimethylacrylamide) (*lower*)

studied to realize non-protein-adsorption and non-cell-adhesion. The development of photoiniferter graft architectures with the PC group located at the growing graft chain was attempted [30, 31]. Figure 20 shows the PC-bearing iniferter-immobilized glass and photopolymerized surface which has hydrophilic polymer graft chain as a spacer and the PC group at the terminal end. The existence of PC groups in polymer grafts was verified by X-ray photoelectron spectroscopy (XPS) and the dye-staining technique [31]. The synergistic effect of the least interaction of the graft chain with proteins and the inherent nature of the PC group was evidenced by the non-cell-adhesion characteristics [31].

5.2
Albuminated Graft Architecture

Because albumin, a major protein in blood, does not activate any body defense systems, an albuminated surface has a nonthrombogenic potential [3]. Surface graft architectures, in which albumin is covalently fixed at

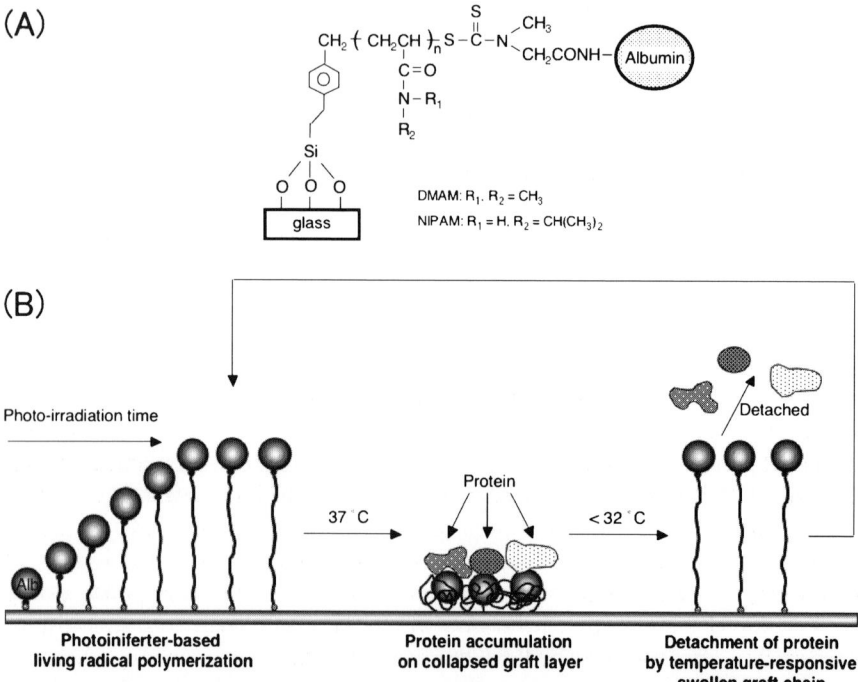

Fig. 21 A Schematics of albumin-bound poly(N-isopropylacrylamide) (PNIPAM)-grafted surface via albuminated DC iniferter. **B** Hypothetical action of temperature-dependent switching of protein desorption by squeezing out and mechanical motion

the growing chain end of hydrophilic polymers such as poly(DMAm) and poly(NIPAM), were prepared by the surface immobilization of a carboxylated iniferter, N-(dithiocarboxy)sarcosine, followed by condensation with albumin and subsequent photoirradiation in the presence of a monomer (Fig. 21) [32]. Surface chemical composition analysis by XPS, immunostaining using a fluorescence-labeled antibody and the measurement of graft thickness, as determined from the $f-d$ curves obtained in water at 25 and 37 °C by AFM, evidenced that the thickness of the graft layer increases with photoirradiation time and albumin molecules exist at growing chain ends. For poly(NIPAM)-grafted surfaces, the interconversion between swollen and collapsed graft chains was observed below and above the LCST of poly(NIPAM). This large structural change (highly swollen versus collapsed state) in the graft chain is beneficial for the self-cleaning of proteins adsorbed on and absorbed in the graft layer. This is due to thermoresponsive on–off of graft-chain conformation. In fact, low-molecular-weight fluorescence-labeled proteins (molecular weight; approximately $5 \times 10^3 \sim 1.5 \times 10^4$) absorbed in the collapsed graft layer was squeezed out by repeated cycles of temperature change, which was visualized by CLSM. The potential application of a thermoresponsive graft with albumin covalently fixed at its growing chain end was discussed in terms of an active nonfouling surface design based on the temperature-dependent switching of phase transition.

6
Surface Derivatization via Cross-Recombination [40]

The simultaneous photolyses of two different species of dithiocarbamates (R_1 – DC and R_2 – DC) in the absence of monomers and at high concentrations result in enforced recombination between alkyl radicals produced by photolysis to produce three types of alkyl–alkyl recombination product namely R_1 – R_1, R_2 – R_2 and R_1 – R_2, as shown in Scheme 5. If the concentration of one species (R_1 – DC) is much higher than the other (R_2 – DC), a relative fraction of heterotopically recombined product, R_1 – R_2, must increase. If such a heterotopic cross-recombination reaction predominantly occurs at a solution/solid interface, an R_2-alkylated dithiocarbamated surface is converted into an R_1-alkylated surface, as schematically shown in Scheme 6.

UV irradiation to a thin solution film (approximately 6 µm in depth) containing a dithiocarbamate substance with an alkyl functional group at a relatively high concentration, which is overlayered on the dithiocarbamated surface, resulted in surface alteration in terms of chemical species and physical properties. These depend on the type of functional group of a dithiocarbamated substance used. Figure 22 lists functional dithiocarbamated substances prepared for this purpose: the functional groups include phenyl, methyl, hydroxyl, carboxyl, phosphonyl, amino,

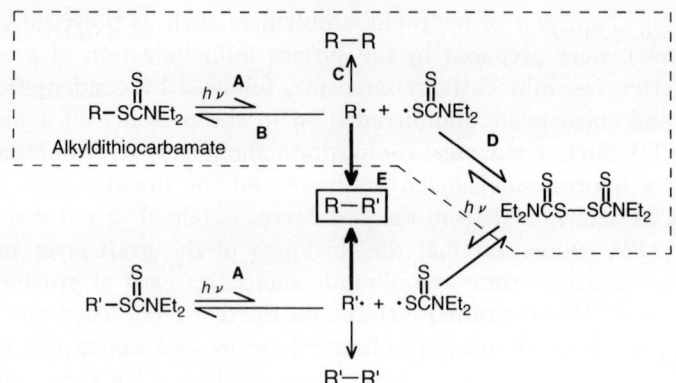

Scheme 5 Cross-combination reactions of two different species of DC derivatives under simultaneous photolysis

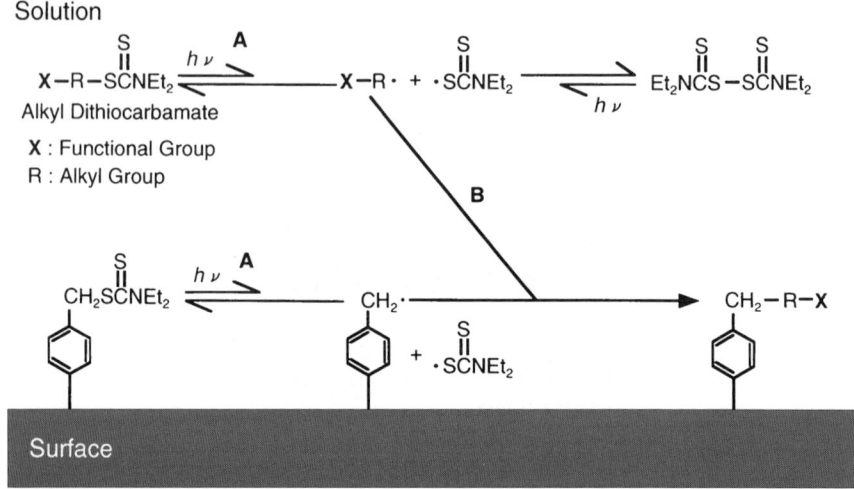

Scheme 6 Surface functional derivatization based on simultaneous photolysis of iniferters in solution and on a surface

dimethylamino, phosphatidylcholine, methoxylated oligo(ethylene glycol) and oligo(N,N-dimethylacrylamide) groups. Using these functional-group-bearing dithiocarbamates, surface derivatization via heterotopic cross-recombination should occur, in principle.

Examples of surface functional derivatization by iniferter-based cross-recombination are shown below. A dithiocarbamated benzyl polymer surface was used as a model surface. When n-propyl N,N-diethyldithiocarbamate was used, XPS measurements showed that the complete loss of both N and S atoms but a marked increase in the C content of the surface was noticed.

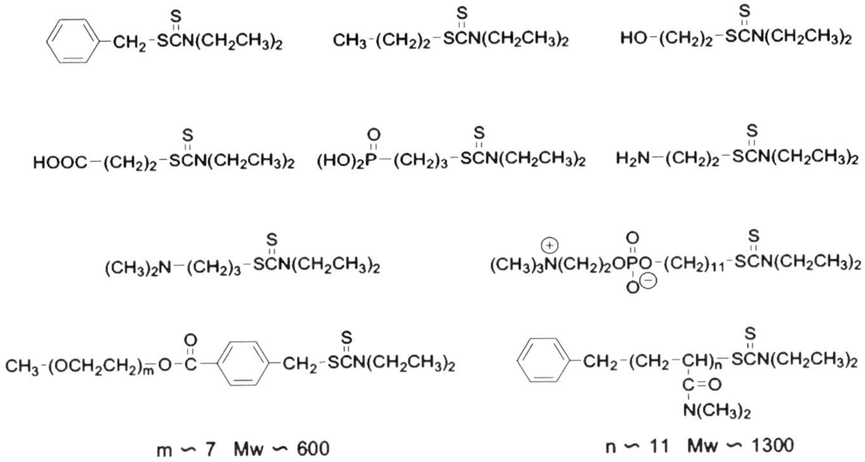

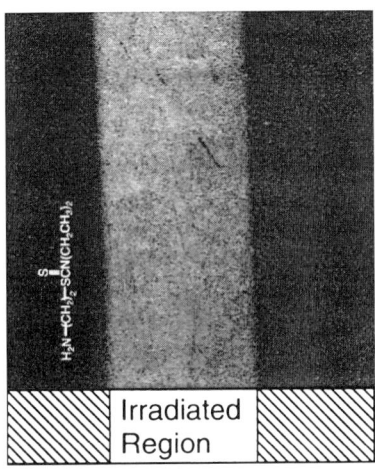

Fig. 22 Various types of DCs with different functional groups

Fig. 23 Surface functional group-exchange via trans-recombination of DCs: Diethyldithiocarbamate-derivatized surface is immersed into aminoethyl diethyldithiocarbamate-containing methanolic solution, followed by photoirradiation. After thorough washing with methanol and immersion in a dye solution, the scanned confocal scanning laser microscopy (CLSM) image showed that only the photoirradiated region was stained, indicating that a surface functional group was exchanged to form an amino group

Concomitantly, the treated surface became water-repellant, indicating that the surface was converted to an ethylstyrenated surface. On the other hand, when 2-aminoethyl dithiocarbamate was used, the treated surface was stained with an anionic dye. Figure 23 showed that only the UV-irradiated portion was stained with a dye. Both dithiocarbamated oligo(ethylene glycol) and oligo(*N*,*N*-dimethylacrylamide) were derivatized on these surfaces, which were verified by wettability (very low water contact angle) and XPS surface analyses (loss of S atoms and increased O content). The carboxyl group was also derivatized using a carboxylated dithiocarbamate, which was verified by chemical coupling with fluorescein under CLSM. Thus, using the inherent nature of iniferter photolysis and the heterotopic recombination feature at a high concentration, a desired functional group can be derivatized on ate photoirradiated surface region. Cross-recombination under appropriate reaction conditions leads to creation of new design tools for surface functionality.

7
Biomedical Applications

The advantages of this surface quasi-living photopolymerization technique are (1) regional control, (2) two-dimensional precision, and (3) self-perpetuating polymerization, enabling chain-length control. A two-dimensional micropattern regionally grafted with different polymers or regionally grafted with different chain lengths can be processed using both a projection mask and controlled sample positioning using a motor-driven sample stage. This enables the construction of a high-throughput tool for the screening and evaluation of cell–material interactions using one sample.

7.1
Micropatterned Tissue Formation

A two-dimensional micropatterned tissue can be easily obtained by utilizing the inherent differences in cell adhesiveness between different micropatterned photografted regions. This was attained by photoiniferter graft polymerization with a projection mask placed on an iniferter-derivatized surface. Since protein adsorption and cell adhesion are markedly suppressed on nonionic graft polymers, such as polyDMAm, any anchorage-dependent cells such as endothelial cell adhere and proliferate only on nonirradiated surfaces, resulting in the formation of a two-dimensional patterned tissue or cellular sheet (Fig. 24).

7.2
Multimicroprocessed Surfaces for Cell Adhesion and Proliferation

The sequential graft polymerization using different monomers at different regions of a surface using a precisely moving step motor with a projection mask, according to Fig. 25, enabled the preparation of a multimicroprocessed surface with unprecedented different photoiniferter-graft-polymerized regions including poly(DMAm), poly(HEMA), poly(sulfopropyl methacrylate)

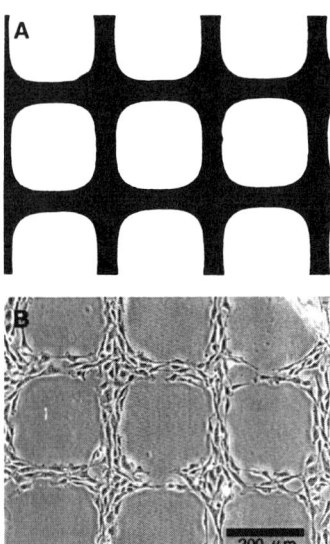

Fig. 24 Micropatterned tissue: photomask used for regional differential non-cell and cell adhesiveness (**A**) and the patterned endothelial tissue (**B**). Photoiniferter polymerization occurs only on photoirradiated regions, which is non-cell-adhesive

[poly(SPMA)], poly(MA) and poly[(N-dimethylamino) propyl acrylamide methiodide] [abbreviated poly(DMAPAAmMeI)]. The adhesion and proliferation of endothelial cells were markedly suppressed on the former two nonionic polymer-grafted regions, whereas the latter two ionic polymer-grafted surfaces promoted cell adhesion and growth. The cell adhesion in the poly(SPMA) region was suppressed with time [22]. In order to produce a large amount of samples in which each sample has a multimicroarray of different natures of graft chains, a laboratory-scale factory driven by a semiautomatic procedure was devised, by which high reproducibility was attained, avoiding batch-to-batch differences [35].

More complex multimicroprocessed gradient surfaces with different graft regions (different photograft-copolymer-thickness gradient surfaces) were prepared according to the procedure shown in Fig. 26 [23]. AFM showed that photoirradiation through a mask in the presence of a monomer solution (DMAAm, DMAPAAmMeI, or MA) under continuous sample movement yielded a graft-polymer-layer thickness gradient surface (Fig. 27). Upon seeding of endothelial cells, cell adhesion potential varied on the surface regions with the type of grafted polymer and graft-chain length. Figure 28 shows cell population–gradient distance relationships. Regardless of graft thickness, cell adhesion was enhanced by the cationic graft polymer poly(DMAPAAmMeI) followed by the anionic graft polymer poly(MA). Among the nonionic graft polymers, poly(DAMAm) was the least cell-adherent surface. The general

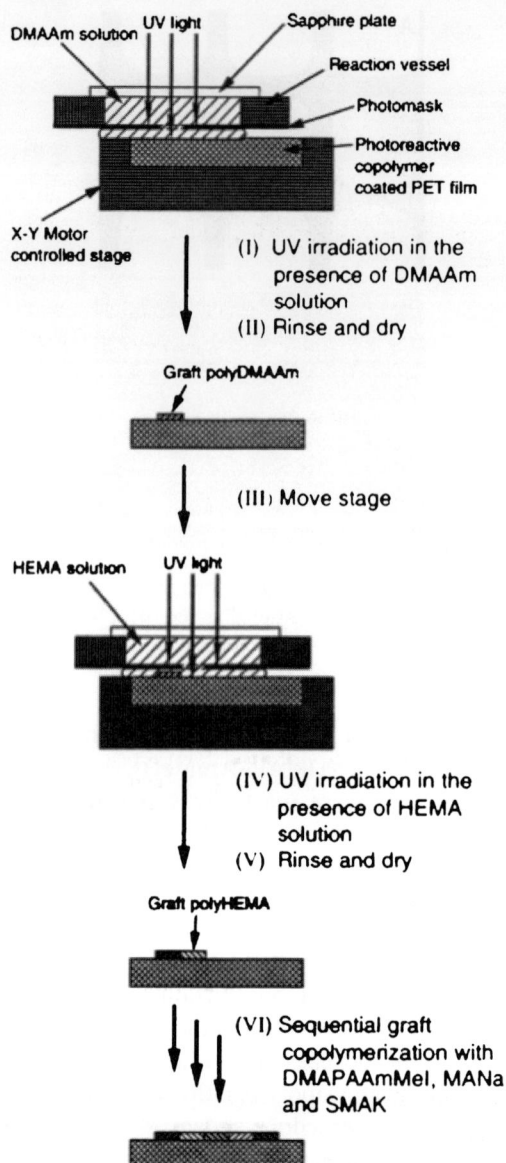

Fig. 25 Schematic diagram of the presentation of a regionally specific micropatterned surface with an unprecedented five different photograft-copolymerized regions using the combination of a photomask and an X–Y step motor-controlled stage

tendency was that an increase in graft thickness appeared to gradually reduce cell population except for the poly(DMAm)-grafted surface, where cell adhesion ceased abruptly above a certain graft thickness. This study shows

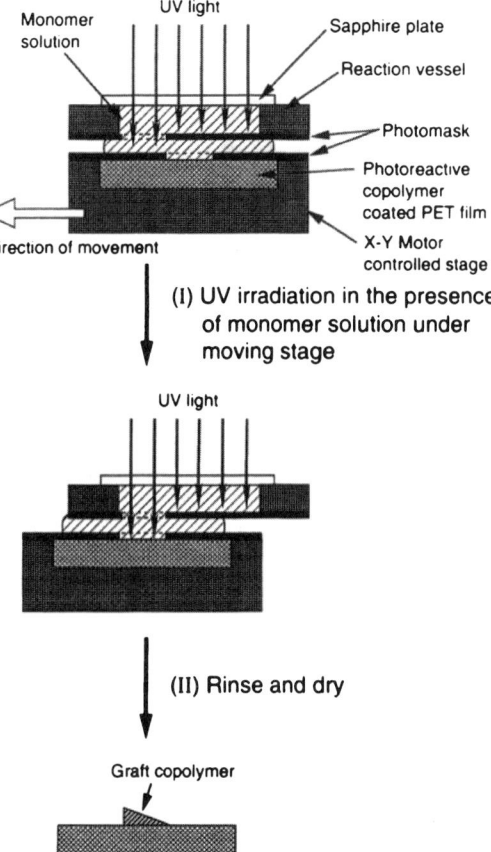

Fig. 26 Schematic diagram of the preparation of a gradient surface varying unidirectionally in thickness of the photograft-copolymerized layer by using the combination of two types of photomasks and the X–Y step motor-controlled stage

that, using this photoiniferter graft copolymerization procedure having the abilities of regional control of both photoirradiation and photoirradiation time, region- and chemical-specific surface modification on the nanoscale thickness and with micron level-regional precision has been achieved, representing a significant advance in the microprocessing of biomedical devices and diagnostic tools [23].

7.3
High-Throughput Screening for Tissue Compatibility

Using the above multimicroprocessed surfaces, which are manufactured by semiautomatic manipulations, early events of tissue–material interactions possibly predicting the tissue compatibility of implant materials have been

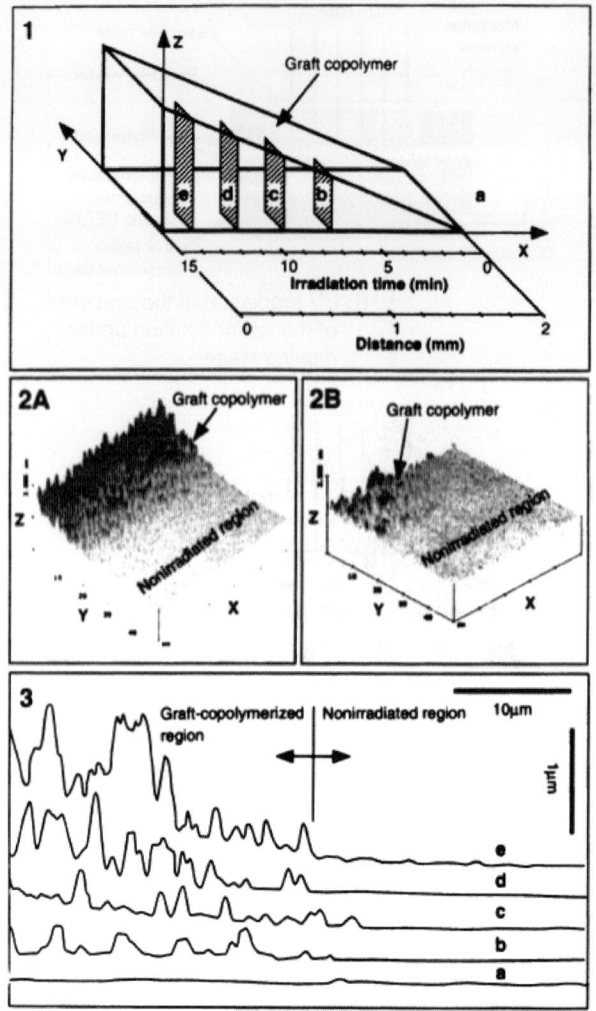

Fig. 27 AFM sampling locations of the grafted region (1) and surface topological features of photograft-copolymerized surfaces as observed by AFM (2). AFM images of the gradient surface regions at 200 μm (position e in 1) (2A) and 1100 μm (position b in 1) (2B), which correspond to irradiation times of 13.5 and 6.75 min, respectively. Line scan spectra for selected regions corresponding to those in part 1 of the gradient film show the measured film thickness

thoroughly studied by the research group of Professor J. Anderson (Case Western Reserve University, Cleveland, USA) in collaboration with the author's group. The following are some results for multimicroprocessed surfaces with different graft polymers.

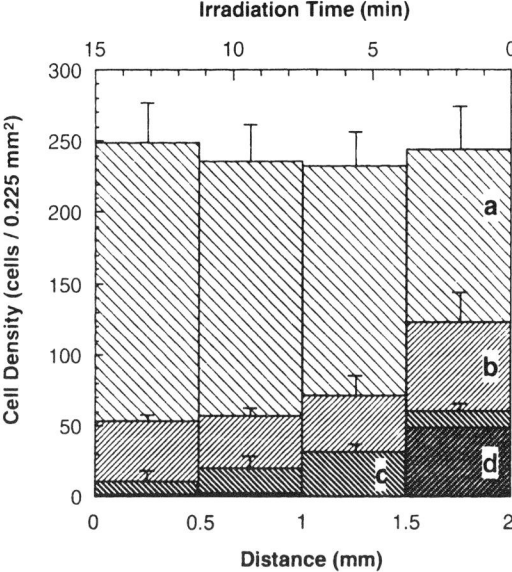

Fig. 28 Cell density versus distance histogram at 3 days after incubation on item* a non-grafted photoreactive copolymer surface, and graft-polymer-layer-thickness gradient surfaces of **a** polyDMAPAAmMeI, **b** polyMA, **c** polyDMAAm. The number of cells per unit area (0.225 mm^2) is plotted against the unit distance (500 μm) from the edge of the photomask

7.3.1
Macrophage and its Fused Foreign-Body Giant Cell

A common component of the foreign-body response to implanted materials in tissues is the presence of adherent macrophages that fuse to form foreign-body giant cells (FBGCs). These multinucleated cells have been shown to concentrate the phagocytic and degradative properties of macrophages at the implant surface and are responsible for the damage and failure of an implant [33]. Therefore, the modulation of the presence or actions of macrophages and FBGCs at the material–tissue interface has been extensively studied recently. A possible means of achieving this is to induce the apoptosis of adherent macrophages, which results in no inflammation. A hypothesis is that the induction of the apoptosis of adherent cells used as biomaterials can be influenced by the chemistry of the surface of adhesion [33]. This is demonstrated by the formation of a multimicroarray of different graft polymers in different regions by the combined use of a photoiniferter technique and a stepping motor-driven photomask. The results show that surfaces displaying hydrophilic and anionic chemistries induce the apoptosis of adherent macrophages at a higher magnitude than hydrophobic or cationic surfaces [36]. Additionally, the level of apoptosis for a given surface is inversely related to a surface's ability to promote the fusion of macrophages to FBGCs. This suggests that macrophages fuse with FBGCs to escape apoptosis. Additionally, the use of hydrophilic and anionic substrates led to decreased rates of both monocyte/macrophage adhesion and its fusion. This was also verified by an in vivo implantation study [36, 37]. These studies clearly demonstrate that biomaterial-adherent cells undergo material-dependent apoptosis in vivo, rendering potentially harmful macrophages nonfunctional while the surrounding environment of the implant remains unaffected [33, 36, 37].

7.3.2
Cytokine Production

IL-8 is a proinflammatory/antiwound healing cytokine (interleukin) whereas IL-10 is antiinflammatory. IL-10 expression level significantly increased in monocyte/macrophage adherent to hydrophilic and anionic surfaces but significantly decreased in cationic-surface-adherent monocytes/macrophages [40]. Conversely, IL-8 expression level significantly decreased in cells adherent to nonionic and anionic surfaces. Further analysis revealed that nonionic and anionic surfaces inhibited monocyte adhesion and macrophage fusion to FBGCs. Therefore, nonionic and anionic surfaces promote an antiinflammatory type of response by inducing selective cytokine production by biomaterial-adherent monocytes and macrophages [39].

This discussion provides crucial information on the design criteria for tissue-compatible surfaces since adherent monocytes/macrophages secrete cytokines in response to an implant that directs the recruitment of more leukocytes (of both myeloid and lymphoid lineages) and other cell types, including fibroblasts, to the tissue–material interface. Thus, material responses to macrophages, including adhesion, fusion to FBGCs, apoptosis and cytokine secretion, can direct the overall response to an implanted biomedical device and prosthesis, determining the fate of implants [33, 36, 37, 39, 40].

8
Conclusions

Iniferter-based surface photopolymerization and surface derivatization have considerable advantages over conventional free-radical polymerization and surface modification techniques. In such photopolymerization and photoderivatization, the surface design, including that of a described chemical species, controlled graft chains, regiospecific addressability and dimensional precision are feasible. Highly spatioresolved grafted surfaces were subjected to qualitative and quantitative analyses using high-resolution CLSM and AFM. Well-designed highly spatioresolved surface microarchitectures served as biocompatible surfaces [20] for the high-throughput screening of substrate-dependent cell adhesion potential [17, 25–28] and as physicochemical interaction model surfaces with proteins [22] and water-swollen layers [23]. On the basis of the logical progression of the quasi-living polymerization approach toward the realization of a repetitive, generational, the branching topology of surface graft chains was elaborated here. The control of macromolecular configuration, shape, pattern, form, geometry and topography is a key axiom in precision surface graft architecture. Thus, the progression of tree-like morphogenesis now becomes a reality. Programmed morphogenesis systems produce diverse multibranching topographies. The ramifications extend far beyond the control of the overall molecular shape to include choices of parameters such as internal and external rigidities, lipophilicity and hydrophilicity, degrees of void volume and excluded volume, density gradient, complementary functionality, and environment cooperativity. The ability to precisely multifunctionalize films in one physical sample with regional specificity enabled the simultaneous evaluations of cellular responses in grafted films, thus efficiently reducing sample-to-sample variability. However, the greater implications of this technique rest in the possibility of precisely controlling surface chemistry to optimize cellular responses on the micrometer order. In addition, graft-polymer-layer-thickness surfaces allow a controlled examination of the effect of polymer chain length on cellular responses in one physical sample. With these novel surfaces, simultaneous characterizations of cellular responses in multiple surface microenvironments were realized.

Acknowledgements This review article is dedicated to Dr. Takayuki Ohtsu (Professor Emeritus, Osaka City University) who pioneered photoiniferter polymerization. A large number of studies by his research group stimulated and directed me to conduct a series of surface microarchitecture studies focusing on biomedical applications. The author also appreciates Professor Rainer Jordan, volume editor of this special issue, who carefully edited this article with patience.

References

1. Ratner BD, Hoffman AF, Schoen FJ, Lemons JE (1996) Biomaterials Science: an Introduction to Materials in Medicine. Academic, New York, p 193
2. Andrade JD (1985) Polymer Surface Dynamics, Vol. 1.Plenum, New York
3. Kim SW, Jacobs H (1996) Blood Purif 14:357–362
4. Ishihara K, Ueda T, Nakabayashi N (1990) Polym J 22:355–368
5. Ikada Y (1994) Biomaterials 15:725–733
6. Otsu T, Matsumoto A (1998) Adv Polym Sci 136:75–137
7. Otsu T, Yoshida M, Tazaki T (1982) Rapid Commun 3:133–40
8. Otsu T, Yaoshida M (1982) Rapid Commun 3:127–32
9. Ishizu K, Khan RA, Ohya Y, Furo M (2004) J Polym Sci A Polym Chem 42:76–82

10. Ishizu K, Katsuhara H, Kawauchi S, Furo M (2004) J Appl Polym Sci 95:413–418
11. Ishizu K, Katsuhara H, Itoya K (2005) J Polym Sci A Polym Chem 43:230–233
12. Bowman NC, Anseth SK, Luo N, Lovell GL, Lu H (2001) Polym Mater Eng 85–156
13. Luov N, Hutchison BJ, Anseth SK, Bowman NC (2002) J Polym Sci A Polym Chem 40:1885–1891
14. Luo N, Hutchison BJ, Anseth SK, Bowman NC (2002) Macromolecules 35:2487–2493
15. Ward HJ, Bashir R, Peppas AN (2001) J Biomed Mater Res 56:351–360
16. Kitano H, Ohhori K (2001) Langmuir 17:1878–1884
17. de Boer B, Simon KH, Werts LPM, van der Vegte WE, Hadziioannou G (2000) Macromolecules 33:349–356
18. Zaremski YM, Chernikova VE, Izmailov GL, Garina SE, Olenin VA (1996) Macromol Rep A 33:237–242
19. Zarenskii YM, Olenin VA (1991) Zh Prikl Khim 64:2145–2149
20. Nakayama Y, Matsuda T (1999) Macromolecules 32:5405–5410
21. Nakayama Y, Matsuda T (1996) Macromolecules 29:8622–8630
22. Nakayama Y, Matsuda T (1999) Langmuir 15:5560–5566
23. Higashi J, Nakayama Y, Marchant RE, Matsuda T (1999) Langmuir 15:2080–2088
24. Lee J, Nakayama Y, Matsuda T (1999) Macromolecules 32:6989–6995
25. Nakayama Y, Sudo M, Uchida K, Matsuda T (2002) Langmuir 18:2601–2606
26. Lee HJ, Matsuda T (1999) J Biomed Mater Res 47:564–567
27. Nakayama Y, Anderson JM, Matsuda T (2000) J Biomed Mater Res 53:584–591
28. Kidoaki S, Nakayama Y, Matsuda T (2001) Langmuir 17:10870–10872
29. Kidoaki S, Ohya S, Nakayama Y, Matsuda T (2001) Langmuir 17:2402–240
30. Matsuda T, Kaneko M, Ge S (2003) Biomaterials 24:4507–4515
31. Matsuda T, Nagase J, Hirano A, Kidoaki S, Nakayama Y (2003) Biomaterials 24:4517–4527
32. Matsuda T, Ohya S (2005) Langmuir 21:9160–9665
33. Brodbeck WG, Patel J, Voskerician G, Christenson E, Shive MS, Nakayama Y, Matsuda T, Ziats NP, Anderson JM (2002) Proc Natl Acad Sci USA 99:10287–10292
34. Idiris A, Kidoaki S, Usui K, Maki T, Suzuki H, Ito M, Aoki M, Hayashizaki Y, Matsuda T (2005) Biomacromolecules 6:2776–2784
35. Nakayama Y, Anderson JM, Matsuda T (2000) J Biomed Mater Res 53:584–591
36. DeFite KM, Colton E, Nakayama Y, Matsuda T, Anderson JM (1999) J Biomed Mater Res 45:148–154
37. Brodbeck WG, Shive MS, Colton E, Nakayama Y, Matsuda T, Anderson JM (2001) J Biomed Mater Res 55:661–668
38. Brodbeck WG, Nakayama Y, Matsuda T, Colton E, Ziats NP, Anderson JM (2002) Cytokine 18:311–319
39. Brodbeck WG, Voskerician G, Ziats NP, Nakayama Y, Matsuda T, Anderson JM (2003) J Biomed Mater Res 64A:320–329
40. Nakamata K (1998) MS Dissertation, Osaka Institute of Technology. Matsuda T, Nakamata K, Hirano J, Nakayama Y (in contribution)

Polymer Brushes by Anionic and Cationic Surface-Initiated Polymerization (SIP)

Rigoberto Advincula

Department of Chemistry and Department of Chemical Engineering,
University of Houston, Houston, TX 77204, USA
radvincula@uh.edu

1	Introduction	108
1.1	Definition	108
1.2	Surface-Initiated Polymerization (SIP)	110
2	Conventional Cationic and Anionic Polymerization	111
2.1	Anionic Polymerization	111
2.2	Cationic Polymerization	112
3	Anionic Surface-Initiated Polymerization	113
3.1	Differences with Conventional Anionic Polymerization	113
3.2	Anionic SIP on Particles	113
3.3	Anionic SIP on Flat Substrate Surfaces	117
3.3.1	Attempts at Anionic SIP	118
3.3.2	Homopolymers and Characterization of Grafted Polymers at Surface	120
3.3.3	Block Copolymers Grafted from SiOx and Au Surfaces	124
4	Cationic Surface-Initiated Polymerization	126
4.1	Cationic SIP on Particles	126
4.2	Cationic Surface-Initiated Polymerization on Flat Surfaces	129
5	Conclusions	132
References		133

Abstract The formation of homopolymer and block copolymer brushes grafted from flat and nanoparticle surfaces via surface-initiated anionic and cationic polymerization methods is reviewed. Unique properties of these chain addition polymerization methods distinguish them from free-radical and living-radical methods, i.e., primarily the formation of charged reactive propagating centers. This involves the use of methods that preserve the reactivity of the charged species, where the monomer, solvent quality, and lack of terminating species allow for grafting to surfaces and for the formation of homopolymer and block copolymers. While these initiators are analogous to solution and bulk methods and adapted to surfaces, their mechanisms do not necessarily follow their counterparts. Several systems for surface-initiated polymerization (SIP) will be reviewed including early attempts at "grafting onto" and "grafting from" particles. For initiation, alkylsilane or alkylthiol anionic initiators are grafted onto planar and particle surfaces by self-assembled monolayer (SAM) techniques. For the cationic (carbocationic) polymerization methods, methods of tethering Lewis acids to surfaces have been reported. The grafted polymer chains can be investigated in situ using a number of surface-sensitive spectroscopic and microscopic techniques. They can also be analyzed ex situ when the polymer chains are removed from the substrate surface. Activation of the grafted initiator, control of polymerization conditions, and removal of excess activators are emphasized. Interesting differences in particle properties, morphology, thickness, grafting density, and polymerization conditions contrast anionic and cationic charged species from other SIP mechanisms. The problems and potential of these techniques will also be discussed. The formation of block copolymer sequences highlights a unique utility of living anionic and cationic polymerization techniques on surfaces.

Keywords Anionic · Cationic · Initiator · Nanoparticles · Surface initiated polymerization

Abbreviations
SIP	Surface-initiated polymerization
SAM	Self-assembled monolayers
MW	Molecular weight
R_g	Radius of gyration
LASIP	Living anionic surface-initiated polymerization
ATRP	Atom-transfer radical polymerization
ROMP	Ring-opening metathesis polymerization
TEMPO	2,2,6,6-tetramethyl-1-piperidyloxy
RAFT	Reversible addition fragmentation chain transfer
NMR	Nuclear magnetic resonance
GPC	Gel permeation chromatography
XPS	X-ray photoelectron spectroscopy
AFM	Atomic force microscopy
TEM	Transmission electron microscopy
PS	Polystyrene
PBd	Polybutadiene
PI	Polyisoprene
DPE	1,1-diphenylethylene
THF	Tetrahydrofuran
QCM	Quartz crystal microbalance

1
Introduction

1.1
Definition

By definition, polymer brushes are made up of polymer chains grafted (tethered) by one end to a surface or an interface (Fig. 1) [1–3]. The density can be small or high; in the latter case, the polymer chains are crowded and forced to stretch in order to avoid other chains. This results eventually in an equilibrium condition where no external field is necessary to force the chains into this geometry.

The physical properties of polymer brushes have long intrigued polymer physicists. Interestingly, a variety of mechanisms, polymerization parame-

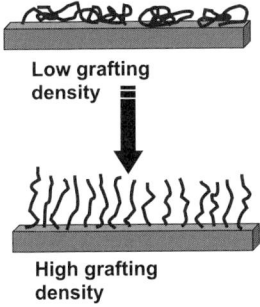

Fig. 1 Polymer brushes with different grafting densities

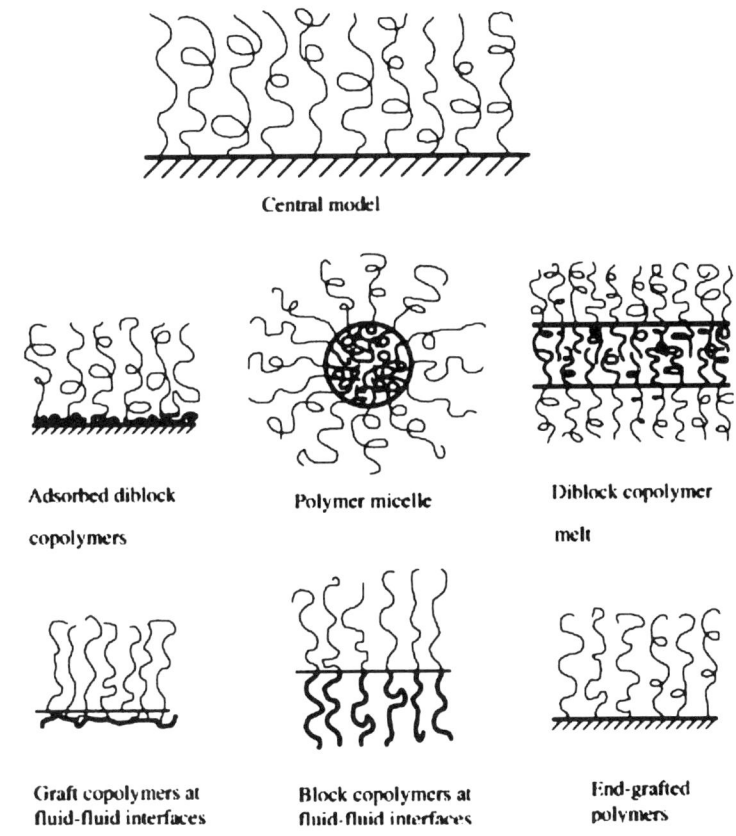

Fig. 2 Polymer brushes at various interfaces and their importance to surface science and colloidal phenomena [8]

ters, and conditions make polymer brushes one of the most important synthetic developments in the field of polymer chemistry. First, a number of assumptions based on polymers in solution or in bulk do not hold as a result of reduced dimensionality (conformations, random-walk, etc.). When the molecular weight (MW), the chain polydispersity, the density of grafting, the homopolymer and graft composition, the chain architecture and block compositions are changed, the properties of these polymers will vary dramatically at the interface. The same can be said when external fields are applied to these systems, e.g., temperature, solvent (solution), pressure, electric fields, magnetic fields, etc.

Thus, fundamentally the interest is in testing the limits and theory of polymer behavior in end-tethered systems, e.g., viscoelastic behavior, wetting and surface energies, adhesion, shear forces relevant to tribology, etc. It should be noted that relevant surfaces and interfaces can also refer to polymers adsorbed in liquid–liquid, liquid–gas, solid–gas, and solid–liquid interfaces, which makes these polymer systems also of prime importance in interfacial science and colloidal phenomena (Fig. 2). Correspondingly, a wide number of potential applications can be enumerated ranging from lubrication and microelectronics to bioimplant surfaces.

1.2
Surface-Initiated Polymerization (SIP)

In general, in the field of materials or condensed matter, the preparation of polymer brushes on solid surfaces is of great interest for surface modification and composite material preparation [4–6]. A number of model surface grafting techniques have been used on planar surfaces and particles and have been the subject of previous reviews. While a number of polymer brush preparation methods have been reported using physisorption or chemisorption or so-called "grafting onto" methods, the emphasis of this review is on surface-initiated polymerization (SIP) methods or "grafting from" methods.

SIP promotes polymerization of monomers directly from initiator sites already attached to surfaces, in which case the activation of initiator and the diffusion of monomers to the reactive sites become primary factors [7]. It is therefore possible to prepare high-density grafted polymers where the average distance between grafting points is much smaller than the radius of gyration (R_g). The surface-initiated *addition polymerization* approach to polymer brushes has been reviewed by others [8]. Polymer brushes can be prepared by a number of polymerization mechanisms including free-radical [9], cationic [10], ring-opening metathesis polymerization (ROMP) [11], atom-transfer free-radical polymerization (ATRP) [12–16], polymerizations using 2,2,6,6-tetramethyl-1-piperidyloxy (TEMPO) [17], anionic polymerization [18], reversible addition fragmentation, and transfer (RAFT) polymerization [19, 20]. All these methods are suitable for polymerizing different types of monomers on a variety of flat surfaces and particles. The focus of this review is on anionic and cationic polymerization methods by SIP.

2
Conventional Cationic and Anionic Polymerization

Before looking at individual anionic and cationic SIP systems, it would be worthwhile to review common knowledge about anionic and cationic polymerization [21]. In general, these are addition polymerizations based on the propagation of an ionic active specie and stabilized by a counterion. The mechanism of polymerization is strongly influenced by the counterions primarily due to solvation effects and temperature. Thus, they are more complex than radical polymerizations but at the same time more versatile. They are very selective in the type of monomers that can be polymerized. Because of their polar nature, substituent groups must be able to stabilize the *carbanions* and *carbocations* that are formed. Also, because of the nature of the ion pair, solvent polarity is very important. Aprotic solvents tend to prevent transfer to solvent and termination. They should be free of electrophilic impurities that can react with ionic sites. They should dissolve both monomers and growing polymers to allow for heterogeneous polymerization. Thus, high-purity solvents are essential. Additives can also be used to influence solvent polarity and to break "aggregates." Finally, if all the conditions are met for propagation such that the use of the right solvent and high-vacuum conditions prevent transfer and termination (disproportionation or coupling), the polymerization will give a "living" polymerization.

Living polymerizations have several advantages: formation of block copolymers by sequential addition of monomers, multifunctional initiators, designed coupling reactions, formation of graft and star copolymers (miktoarms) with multifunctional polymers or difunctional monomers. Several tests for living polymerizations will show that (1) a polymer has a linear relationship for Mn and percent conversion, (2) the slope of the log (ln) relationship of the initial monomer and the remaining monomer ratio and the time of polymerization gives the rate constant of propagation, (3) the polymer is able to grow with further addition of monomers or is capable of forming block or graft copolymers, (4) the polydispersity ratio is narrow, approaching one, and gives a Poisson distribution with respect to the monomer concentration and the MW (degree of polymerization) at 100% yield.

2.1
Anionic Polymerization

Although cationic and radical polymerizations are now capable of living polymerizations, anionic living polymerizations are the most well-studied polymerization systems. However, it cannot be said that *all* anionic polymerizations are living or controlled. The salient aspects of the anionic polymerization mechanism are as follows. (1) The propagating chain is an anion: initiation is brought about by species that undergo nucleophilic addition to the monomer. (2) Monomers have substituents capable of stabilizing the carbanion through resonance or induction, e.g., nitro, cyano, carboxyl, vinyl, phenyl. (3) The strength of a base necessary to initiate polymerization depends on the monomer structure, e.g., superglue: cyano acrylates are anionically polymerized by moisture. (4) Most applications of anionic polymerization are initiated by alkali metals and their compounds. (5) Kinetics and mechanisms are better understood compared to cationic polymerizations. (6) Chain transfer may occur in some cases but rarely. (7) Living polymerizations are often observed. In fact, the importance or superiority of anionic polymerizations compared to living cationic and radical polymerizations is that they have been used as the best method of preparing complex macromolecular architectures both in terms of compositions (block copolymers) and architectures (grafts, stars, and miktoarms) because of the nature of living polymerizations [22].

2.2
Cationic Polymerization

Compared to anionic polymerizations, cationic polymerizations are not as well studied and have not been made suitable for complex macromolecular synthesis [21]. Some of the salient features are as follows. (1) The propagating species is a carbocation. Initiation is brought about by the addition of an electrophile to a monomer molecule. (2) Initiators of cationic polymerizations include H_2SO_4, H_3PO_4, $AlCl_3$, BF_3, $TiCl_4$. With Lewis acids, a proton source is required. In addition, various initiators can also bring about cationic polymerizations: triphenyl metal halides, tropylium halides, iodine, radical cations, etc. (3) Monomer reactivity: the addition of electrophilic species follows Markovnikov's rule: a more stable carbocation intermediate is formed. The double bond is a nucleophile and attacks the electrophilic carbocation reactive center. Also notable is the presence of electron-donating groups, e.g., isobutylene, parasubstituted styrenes, vinyl ethers, etc. (4) Propagation steps are favored by increasing carbocation stability. (5) Solvent polarity influences the polymerization rate: solvation of the ion pair; the more intimate, the lower the propagation rate. (6) Chain-transfer reactions and chain branching are also common in cationic polymerization. Rearrangements are possible and lead to more stable carbocations. (7) Termination is harder to define (related to chain transfer), i.e., usually by the combination of the chain end with counterions. (8) Most cationic polymerizations show first-order rate dependence and the MW in cationic polymerizations is independent of initiator concentration. (9) For cationic copolymerization, the same copolymer equations can be used as in free-radical initiation. The reactivity ratios vary with initiator type and solvent polarity. There is no apparent tendency for alternating copolymers to form, but homopolymer blends and block copolymers are likely. (10) Lastly, there are no observable trends with temperature.

3
Anionic Surface-Initiated Polymerization

3.1
Differences with Conventional Anionic Polymerization

Among the two ionic polymerization techniques mentioned above, a living anionic polymerization should show the best possible control of polymer architecture and composition. Monodispersed homopolymers, complex-block, graft, star, and miktoarm architectures have been accessible primarily by anionic polymerization methods [22]. They have been used to grow polymer brushes from various small particles such as silica gels graphite,carbon black, and flat surfaces [23–26]. Recent results have been reported on living anionic polymerizations on clay [27] and silica nanoparticles [28, 29].

Before reviewing the SIP methods any further, we need to focus again on some specific cases that might be encountered when anionic polymerization is adapted to surfaces. To facilitate control in anionic SIPs, this will require the right choice of initiator, monomer, and solvent. This includes the use of high-purity and anhydrous reagents and an inert atmosphere (often high-vacuum conditions). For MW control, this will require the use of reactive and efficient initiators. Several cases can be cited. Analogous to that of solutions, surface-bound n-alkyllithiums should be some of the most effective initiators. Phenyllithiums are relatively inefficient and unreactive even in polar media and insoluble in hydrocarbon solution and therefore may have poor reactivity. However, in surfaces, they have very different reactivity [26]. On the other hand, $tert$-butyllithium (t-BuLi) has been shown to be an inefficient initiator for styrene polymerization [30, 31]. With solvents, anionic grafting reactions in tetrahydrofuran (THF) and toluene are not well controlled, i.e., most organolithium compounds are not stable in the presence of THF. The use of toluene as a solvent can also lead to chain-transfer reactions [32, 33]. Benzene for most purposes appears to be the most suitable but is not ideal because of toxicity. Finally, with very polar vinyl monomers, the use of alkyllithium initiators will not result in a living polymerization.

3.2
Anionic SIP on Particles

Ideally, SIP from nanoparticles proceeds from surface-attached initiators (Fig. 3) [1, 2]. Early reports involving "grafting onto" via living n-BuLi-anion-initiated polystyrene chains were first described in the 1970s [35, 36]. This usually involved reactive oxide or chlorinated SiOx with the carbanions of the living polymers of polystyrene or polyvinylpyridine. Kinetics and surface coverage were monitored as a function of time and temperature and examined by gravimetric methods. The most abundant "grafting from" anionic SIP reactions were reported on surface-modified particles: carbon black (whiskers, fibers, powders), inorganic oxides of alumina, and silica and titania surfaces (particles) were the primary particle substrates. In particular, numerous reports based on work by Tsubokawa et al. has been reported on various particles and anionic polymerization/initiator systems [37]. These systems are versatile and technologically relevant, but the polymerizations

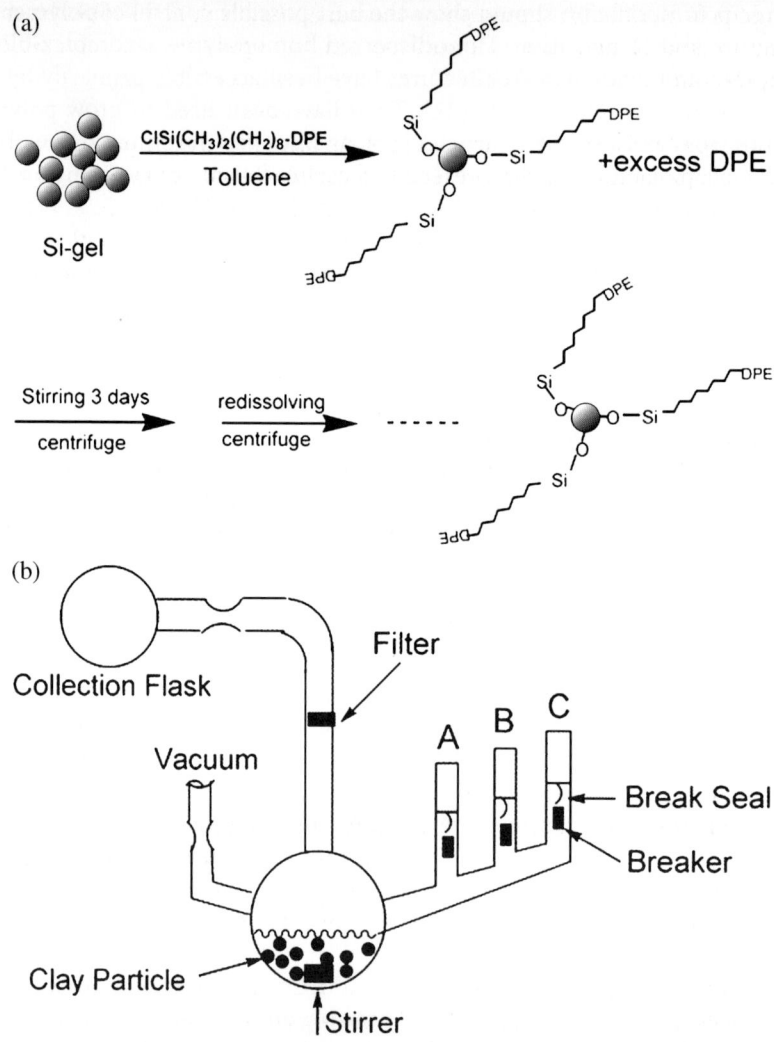

Fig. 3 a Immobilization of DPE initiator precursor and activation by addition of *n*-BuLi. Monomer is eventually added and results in polymerization. **b** Specially made high-vacuum reactor with filter for handling LASIP of nanoparticles. The *scheme* shows the *A* ampule containing styrene, *B* ampule containing *n*-BuLi, *C* ampule containing MeOH [28, 29]

are not necessarily living. Most of these systems in fact involved ring-opening polymerizations initiated on surfaces.

The primary methods for analysis were usually gravimetric, thermal, and spectroscopic in nature but not necessarily correlated with in situ analysis (XPS, AFM, TEM, etc.) or ex situ analysis of surface-bound polymers by degrafting (NMR, MW, polydispersity, etc.). Colloidal stability and homogeneity of the grafting process is a primary concern. A range of these systems were analogous to what has been done in solution and in bulk and should be thoroughly examined in terms of chemistry on flat substrate surfaces. Several examples follow.

(1) Grafting of polyesters on carbon black surfaces by anionic polymerization of β-propiolactone in the presence of carbon black containing an alkali metal carboxylate (CO_2M; M = Li, Na, K, Rb, or Cs) group as catalyst. The catalyst was prepared usually by reacting CO_2H groups on the carbon black surface with an alkali metal hydroxide, which initiated the anionic ring-opening polymerization [39].

(2) In the presence of carbon black containing amino groups, the anionic ring-opening polymerization of γ-Me L-glutamate N-carboxyanhydride (NCA) was initiated to give polypeptide-grafted carbon black. The introduction of amino, methylamino, or dimethylamino groups onto a carbon black surface was usually achieved by the reaction of carbon black containing acyl azide groups with corresponding ethylenediamines [40].

(3) Potassium carboxylate groups introduced onto the surface of carbon fibers initiated anionic polymerization of epoxides (e.g., styrene oxide, epichlorohydrin, and glycidyl phenyl ethers) and cyclic acid anhydrides (e.g., maleic anhydride, succinic anhydride, and phthalic anhydride) in the presence of 18-crown-6 [41].

(4) The grafting of polyesters onto ultrafine inorganic particles, such as SiOx, TiOx, and Ni–Zn ferrite, by the use of COO^-K^+ groups introduced onto the surface has also been investigated. The introduction of COO^-K^+ groups onto the surface was achieved by neutralization of acid anhydride groups, which were introduced by the reaction of 4-trimethoxysilyltetrahydrophthalic anhydride with the hydroxy groups on the surface. The anionic ring-opening copolymerization of epoxides with cyclic acid anhydrides was also reported [42, 43].

(5) The anionic graft polymerization of vinyl monomers onto carbon fiber or graphite powder initiated by metalized carbon fiber has been investigated. The metalation of polycondensed aromatic rings of a carbon fiber surface was achieved by treatment of the carbon fiber with BuLi in N,N,N',N'-tetramethylethylenediamine. The anionic polymerization of methylmethacrylate and styrene was reported. No grafting was observed when carbon fiber was treated simply with BuLi in THF or toluene [44, 45].

(6) Direct anionic polymerization of methylmethacrylate monomers initiated by n-BuLi in the presence of carbon whiskers using crown ethers has been reported. PMMA was grafted onto the surface depending on the propagation of O – Li groups, which were produced by the reaction of O-containing groups on the surface with n-BuLi [46].

(7) Grafting of bisphenol-A polycarbonate on carbon fibers with n-BuLi in the presence of N,N,N',N'-tetramethylethylenediamine and 18-crown-6 has been reported by Drzal et al. [47].

In the case of silica and silicalike particles, Schomaker et al. [48] immobilized a methylmethacrylate (MMA) derivative onto silica (Aerosil) particles, initiated polymerization using a Grignard reagent, and added additional MMA.

Recently, Advincula et al. reported the synthesis of polymer brushes on silica and clay nanoparticles using 1,1-diphenylethylene initiators for anionic SIP (Fig. 3) [27–29]. We recently reported the "living" nature of surface-bound DPE for anionic polymerization on nanoparticles by demonstrating a linear relationship between monomer concentration and MW and by the appearance/disappearance of a red-colored Li-DPE anion complex. In order to demonstrate the living anionic surface-initiated polymerization (LASIP) on silica nanoparticles, DPE was functionalized with alkyldimethylchlorosilane and grafted onto silica particle surfaces. n-BuLi was used to activate the DPE, which allowed the anionic polymerization of the styrene monomer to proceed in benzene solution. A specialized high-vacuum reactor was used to allow polymerization to occur from the surface of the dispersed colloidal particles under anhydrous solution conditions. After addition of the n-BuLi, the dispersion of the DPE functionalized silica particles showed a distinct red color, indicating the formation of an activated nanoparticle-DPE-n-BuLi complex. The degree and mechanism of polymerization were determined by characterizing the grafted and detached polystyrene chains using thermogravimetric analysis (TGA), size exclusion chromatography (SEC), NMR, and FT–IR spectroscopy. In addition, atomic force microscopy (AFM) and X-ray photoelectron spectroscopy were used to characterize the polymer-coated nanoparticles. The importance of activation of the grafted initiator, control of aggregation properties, and removal of the excess n-BuLi for high-MW formation was emphasized. While the polydispersities were broader compared to those obtained by solution polymerization of a free initiator, a living anionic polymerization mechanism was still observed.

In the case of LASIP with clay nanoparticles, polystyrene was grafted using a DPE coinitiator. The montmorillonite clay surface and intergallery interfaces were intercalated with 1,1-diphenylethylene (DPE) modified to be an organic cation as shown in Fig. 4. Its intercalation was confirmed by a series of characterization methods including X-ray diffraction (XRD), FT–IR spectroscopy, TGA, and XPS. The results showed a complete replacement of

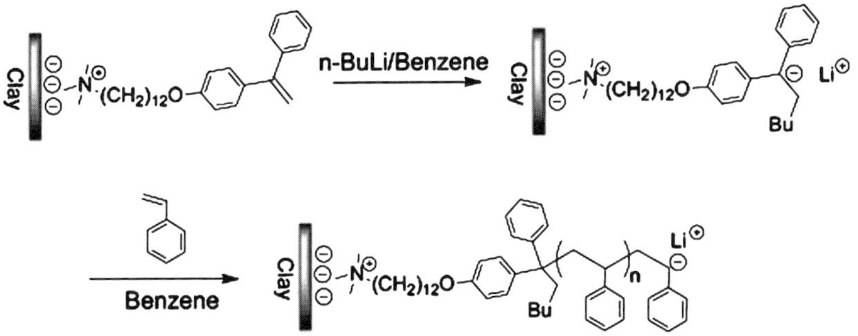

Fig. 4 Schematic illustration of living anionic surface-initiated polymerization (LASIP) from clay surfaces showing initiator design and electrostatic adsorption to the surface of clay [28, 29]

the Na^+ counterions by the charged initiators. LASIP was also performed in a high-vacuum reaction setup. A key challenge was the removal of adventitious moisture, which could terminate the formation of "living anions." This was facilitated by complete drying of the clay prior to polymerization. Different styrene monomer/initiator ratios were used. A living anionic polymerization mechanism was determined from MW data and the MW distribution. A comparison of FT–IR, TGA, XPS, XRD, and AFM data before and after the LASIP confirmed that polystyrene was indeed "grafted from" clay surfaces for these composite materials. The initiation efficiency was also distinguished between surface- and intergallery-interface-bound initiators.

3.3
Anionic SIP on Flat Substrate Surfaces

An important missing ingredient in anionic SIP was also supplied by the use of flat substrates for investigations. Although polymerization on particle surfaces is one of the best ways of obtaining more polymer samples for analysis (because of the high surface-to-volume ratio), polymerization on flat surfaces has advantages primarily due to the availability of surface-sensitive analytical techniques. Also, there is the potential for polymer brush applications in surface modification and patterning of flat substrates [49]. Theoretical predictions have been used to calculate MW and polydispersity of polymer brushes grown by SIP on flat surfaces [50–52]. Other theoretical predictions have been made on conformation and dynamic behavior of tethered polymers at surfaces [53]. There is a need to verify theoretical predictions on block copolymer brush behavior with respect to the Flory–Huggins (χ) interaction parameter, Kuhn length, block volume fraction, and substrate surface energies [54–57]. Interesting mesophases of block copolymer brushes on surfaces have also been predicted based on a variety of surface interactions [58, 59]. Thus, the development of techniques for the synthesis of tethered homopolymer and block copolymer brushes has to be carefully explored in the context of anionic polymerizations on flat surfaces.

The ideal systems for creating suitable initiation sites on surfaces involve the use of SAMs, especially those of *end*-functionalized organic thiols on gold or alysilanes on SiOx surfaces [60]. There are several advantages. First, several ideal surfaces can be prepared that are chemically homogeneous, easy to clean, and virtually free of contaminations, and their roughness, crystallinity, and surface chemistry can be determined a priori. Second, they allow a broad variety of surface-sensitive spectroscopic and microscopic analytical techniques, frequently used for the characterization of thin films. Third, SAMs of different types of functionalized thiols and silanes, for example aryl or *n*-alkyl derivatives, enable the control of the reactivity of the initiator, crucial for obtaining a suitable ratio of the initiation and propagation step of consecutive polymerization reactions. Finally, mixed SAMs allow the control of the lateral concentration and, hence, grafting density of polymer chains.

3.3.1
Attempts at Anionic SIP

In an early work, Oosterling et al. attached chlorosilane-functionalized styrene groups onto silica surfaces (flat and spherical) and initiated polymerization by activating the styrene units using *tert*-butyllithium [61]. The SIP of homopolymers and block copolymers of styrene, isoprene, methylmethacrylate, and 2-vinylpyridine was demonstrated. Polymers were removed from the surfaces using HF solution, allowing the MW and MWD of the chains to be analyzed. The polymers produced had a higher MW and much broader than expected MWD (about 1.3). They reported that anionic polymerization onto flat surfaces was more difficult, even under high-vacuum conditions, instead of the inert atmosphere conditions that were used. Via sequential monomer addition, block copolymers were also grown from microparticulate silica; but efforts to grow graft copolymers from flat silica surfaces failed. It is believed that a major limitation of this work was the use of *tert*-butyllithium (*t*-BuLi) as an initiator for styrene SIP in toluene. *t*-BuLi initiation is very slow in nonpolar solvents, yielding broad MW distributions and higher than expected MWs [62]. Futhermore, sixfold excess of *t*-BuLi had to be used, and the MWs of grafted and nongrafted polymers were assumed to be the same. Lastly, self-polymerization was inevitable due to the presence of surface-bound monomers in the system. In this case, differences between free polymerization in solution and confined polymerization on surfaces were not easily distinguishable.

In the case of work on phenyllithium anionic initiators, the first reported work by Jordan et al. involved the use of a SAM of biphenyllithium moieties on gold substrates to initiate anionic polymerization of styrene [26]. The thickness of the resulting dry polystyrene brush, as estimated by ellipsometry and AFM, was 18 nm. A smooth, homogeneous polymer surface was observed up to the microscopic scale, with a roughness of 0.3–0.5 nm (rms). On the basis of results from in situ swelling experiments, monitored by ellipsometry, a polymerization degree of $N = 382$ and a grafting density of approximately 7–8 chains/Rg^2, or 3.2–3.6 nm^2/chain, were calculated with the use of mean-field theory. Polarized external reflection FTIR spectra of the grafted layer confirmed highly stretched preferentially oriented polystyrene chains.

Independently, the Advincula and Quirk groups have described the use of SAMs of DPE functional groups as coinitiating sites for anionic preparation of polymer brushes on flat substrate surfaces [23–25, 63–67]. These functional groups allow for both "grafting onto" and "grafting from" methodologies based on the versatile chemistry of DPE functional groups for anionic polymerizations in solution [68]. The design parameters for this type of initiator combines DPE functionality, spacers (alkyl chain), and reactive end groups (chlorosilane or thiol) as shown in Fig. 5. The DPE is separated from the silyl or thiol group by an alkyl spacer. The advantage of using DPE is that it can react quantitatively with simple alkyllithiums to form a monoaddition product, a 1,1-diphenylalkyllithium initiating species (a complex ion) [68]. The detailed synthesis procedure for the DPE initiators has been reported [69].

The absence of homopolymerization due to steric hindrance makes DPE chemistry very useful for a surface-bound monolayer since this coinitiator will not self-polymerize on a surface as would surface-bound styrene moieties or other monomers grafted to surfaces [61]. These complexes are usually very reactive and are useful initiators for polymerization of styrenes and dienes in organic solvents. They can also initiate and control anionic polymerization of (meth)acrylates and vinylpyridines at low temperatures in polar solvents such as THF [68]. The two groups described the use of DPE SAM monolayers on flat silica surfaces with *n*-butyllithium as initiators in benzene for the preparation of hydroxy-functionalized polyisoprene brushes by functionalization of the resulting poly(isoprenyl)lithium chain end with ethylene oxide and for styrene polymerization [23–25, 63–67]. The Quirk group utilized a living "grafting to" procedure to prepare tethered polystyrene chains by addition of poly(styryl)lithium to a DPE SAM on silica [63, 64]. Polystyrene brushes were reported using either *n*-butyllithium or *sec*-butyllithium in the presence of THF, which sometimes resulted in limited stabilities of the organolithium compounds [70]. For comparison, the Quirk group grafted poly(isoprenyl)lithium and telechelic polyisoprene polymers onto a DPE monolayer via silyl chloride surface linking chemistry [71].

A variety of surface-sensitive spectroscopic and microscopic methods were critical in the investigation of these systems. In the work by Advincula et al., the composition, thickness, physical and thermal properties, and morphology of the tethered polymer brushes were carefully analyzed [72]. A variety of surface-sensitive techniques such as ellipsometry, contact angle measurements, AFM, quartz crystal microbalance (QCM), FT–IR grazing incidence

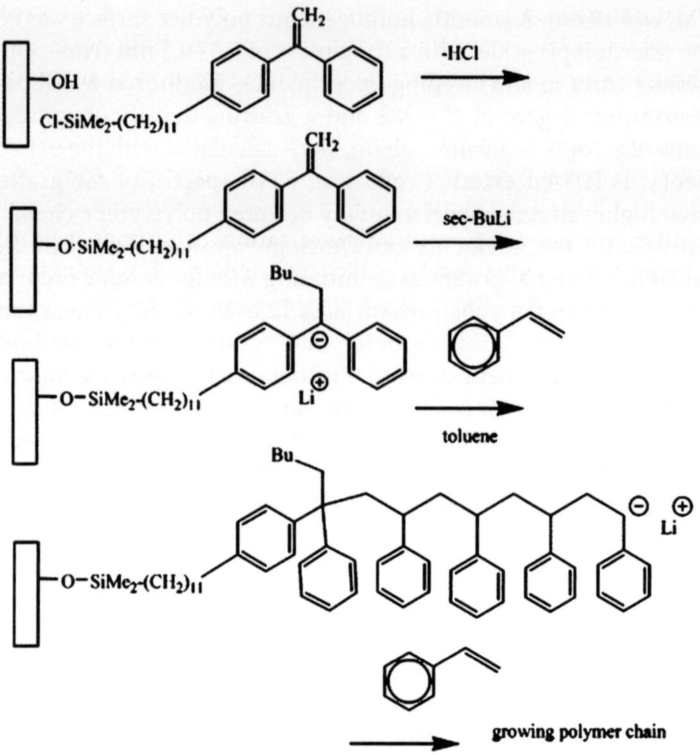

Fig. 5 SAM formation of initiator is essential. Immobilization of DPE initiator followed by polymerization of styrene homopolymer to form PS brushes by SIP [72]

and PM–IRRAS, surface plasmon spectroscopy (SPS), and XPS were utilized. Interesting insights on living anionic polymerization phenomena and mechanisms on flat surfaces were obtained. This was complemented by work on XPS and X-ray reflectivity studies by Quirk et al. [71].

In all these anionic polymerization procedures, high-purity solvent and monomers (anhydrous) are necessary for polymerization [73, 74]. For the SAMs of the initiators, a well-packed monolayer provides a "protective layer" that may prevent reactive groups, e.g., excess n-BuLi, from cleaving the tethered initiator from the surface. For Si surfaces, a well-packed monolayer can eliminate the presence of Si–OH groups that can interfere with anionic polymerization [68, 75, 76].

3.3.2
Homopolymers and Characterization of Grafted Polymers at Surface

Using the polymerization procedure outlined above, polystyrene (**PS**) homopolymers were prepared as shown in Fig. 5 [74]. Polymerization proceeded by introducing n-Bu-Li, removal of the excess, and followed by the addition of the monomer. After polymerization, all films were washed for more than 36 h by Soxhlet extraction in toluene prior to characterization of the surfaces.

The data in Table 1 indicate several attempts made at **PS** polymerization under various conditions on Si wafers and Au substrate. Immediately noticeable is the large distribution of film thickness and contact angles even with similar concentrations and polymerization conditions, e.g., cosolvent and reaction time. These results are comparable to those obtained by Ul-

Table 1 Polymerization under various conditions on Si wafers and Au substrate (from [72])

SAMPLE (polymerization time and $g_{styrene}$)	Activation	Additive (1–2 mL)	Thickness* (nm)	Contact Angle
1 (1 d, 5.1 g)	s-BuLi	none	6.8	95
2 (3 d, 5.0 g)	n-BuLi	THF	10.8 (16%)	70
			16.1 (100%)	76
3 (5 d, 15.1 g)	n-BuLi	THF	13.4 (16%)	95
			13.7 (100%)	89
4 (3 d, 12.0 g)	s-BuLi	BuOli	3.8 (16%)	50
			3.9 (100%)	72
5 (5 d, 12.0 g)	n-Buli	THF	6.3 (16%)	61
			8.2 (100%)	64
6[a,c] (3 d, 7.0 g)	n-BuLi	THF	8.9	82
7[a] (3 d, 2.5 g)	n-BuLi	THF	4.7	74
8 (5 d, 2.0 g)	s-BuLi	none	6.2	81
9[d] (5 d, 2.0 g)	n-BuLi	THF	10.1	80
10 (3 d, 2.5 g)	s-BuLi	THF	6.5	63
11[b] (5 d, 10.9 g)	s-BuLi	TMEDA	23.4[#]	94

All polymerizations were done at room temperature, 24–25 °C. The volume of benzene solvent is 100 mL. Termination was done by the addition of 2 mL methanol.
* Brackets indicate DPE initiator with alkyldimethylchlorosilane solution (0.0001 M) for SAM.
[a] The polymer films of samples 6 and 7 were synthesized under the drybox (inert atmosphere).
[b] For PS homopolymer: 10.9 g styrene, 1.5 mL TMEDA, 5×10^{-3} moles s-BuLi. The substrate used was Au-coated glass.
[#] Thickness obtained by surface plasmon spectroscopy. Almost no difference in thickness was observed with Soxhlet extraction (23.4) and washing procedure (22.8).
[c] Free polymer from solution $M_n = 1.10 \times 10^6$, P.D. = 1.44, by SEC.
[d] Free polymer from solution $M_n = 1.08 \times 10^4$, P.D. = 1.24, by SEC.

man et al. (18±0.2 nm) on Au surfaces and by Quirk et al. [24, 25] using an analogous initiator (9.5±1.2 nm) [26]. Kunz et al. reported thicknesses up to 245 nm on bulk-polymerized polyacrylonitrile brushes using lithium di-*tert*-butyl biphenyl on 3-bromopropylsilane SAMs in the presence of 12-crown-4 [24, 25].

AFM micrographs revealed morphologies uncharacteristic of previously reported grafted **PS** systems using other types of initiators, in particular in the work by Prucker and Rühe using free-radical SIP where they obtained smooth homogeneous and thick films [77]. The morphology is likewise different for the thicker film of **PS** in Au/TMEDA (23.4 nm with 1.2-rms roughness) as shown in Fig. 6. Although the morphology can be explicitly interpreted, a typically lower grafting density is clear. It can result from a different anionic polymerization mechanism, incomplete initiation, or domain formation after postpolymerization (Soxhlet extraction or solvent washing) treatment. One possibility is removal of "unreacted" initiators after Soxhlet extraction or washing. Another possibility is removal of the silane initiators by *n*-BuLi during activation of the DPE. And perhaps another possibility is the high polarity of the "anionic surface" or the interfacial energy involved, affecting the mechanism of the polymerization. These issues remained unresolved and point to the complexity of the polymerization mechanism. Some of these detailed studies were addressed by the two groups, which included the use of polar additives and differentiation from Au- and Si-wafer substrates. Again, the use of XPS, AFM, ellipsometry, contact angle measurements, etc. was critical in this work.

Some interesting hypotheses should point to the uniqueness of doing anionic polymerization on surfaces. It should be noted that the changing interfacial properties of propagating living anions increase the polarity and ionophilicity of a growing "polymerization front" at the interface. Note that the interface of a flat surface is composed of a high concentration of growing chain ends confined in a quasi two-dimensional area. A large difference in surface tension can prevent the flux of monomer (typically nonpolar) at surfaces creating a more heterogeneous interfacial polymerization mechanism. The role of solvent molecules and solvent cage on the propagating anion and Li counterion should be further investigated [78]. A basic assumption about the theoretical treatment by Wittmer was that many simultaneously growing "living" chains compete for the small influx of monomers to the surface [79]. In this case, the increasing difference between polarities at the interface (ionic) and the nonpolar solution subphase may prevent a homogeneous monomer influx or Fickian diffusion. In effect, the polymerization becomes a self-limiting heterogeneous system preventing formation of inherently high-MW brushes. Also, the presence of free initiators in solution (unremoved excess *n*-BuLi or desorbed initiator) competes with the polymerization of monomers, leaving fewer monomers available for polymerization at the interface and hence lower-MW brushes. This explanation seems to sup-

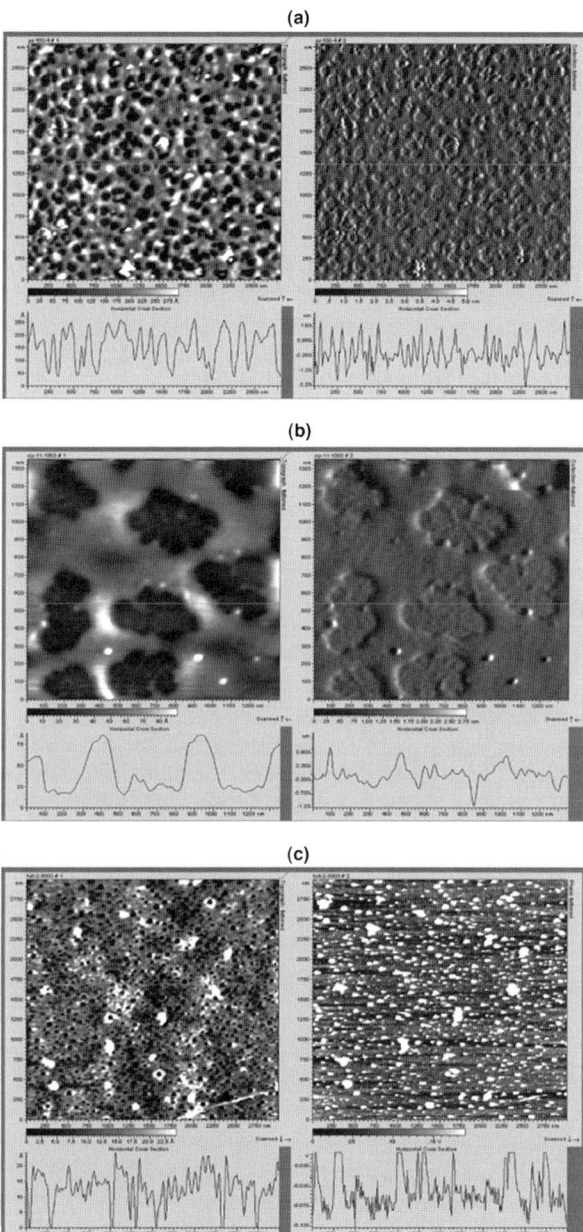

Fig. 6 Morphological AFM images of PS brushes. The topological or height image (*left*) and the amplitude or phase image (*right*) are shown: **a** after polymerization to form PS brushes in benzene/THF (sample 3), X-Y (3.0 × 3.0 μm), and Z (27.5 nm); **b** small-area scan of PS brushes in benzene/THF (sample 3), X-Y (1.35 × 1.35 μm), and Z (8.5 nm), showing dendritic morphology; and **c** after polymerization to form PS brushes in TMEDA on Au-coated glass, X-Y (3.0 × 3.0 μm), and Z (2.25 nm) [72]

port a general trend toward low polymer brush thickness (low MW) with anionic polymerization procedures as reported by others. Interestingly enough, the use of a more polar monomer and a crown ether (12-crown-4) resulted in much thicker films [25]. Important differences should be drawn between anion free systems such as free-radical, ATRP, TEMPO, etc., where polymer brushes are observed to be much thicker. These SIP protocols do not have ionophilic active centers in their mechanism for propagation. There should be some interesting parallel insights to be gained by a careful investigation of analogous surface-initiated cationic polymerization.

3.3.3
Block Copolymers Grafted from SiOx and Au Surfaces

By also using the LASIP procedure, grafted **PS-b-PI** and **PBd-b-PS** block copolymers have been prepared (Fig. 7) [72]. Using silane and thiol–DPE initiators, polymerization was carried out on the SiOx and Au surface by sequential addition of monomers. Typically, after allowing this first reaction to reach completion, the second monomer was added to the living anion, and polymerization of the second block was allowed to proceed. The polymerization was also investigated by SPS [80], AFM, ellipsometry, FT–IR, and XPS. The schematic diagram for the reaction on Au surfaces and the formation of the block copolymers is shown in Fig. 6. The results are summarized in Table 2.

Like the homopolymers, fairly thin polymer thicknesses were observed for the copolymers [72]. Again for these block copolymer brushes, the MW and PDI are unknown, much less the actual constitution and volume of the blocks on the polymer brush. The "presence" of the blocks can be confirmed where the sequence of the polymerization procedure assures consumption

Table 2 Summary of thickness and contact angle measurement results for block copolymers [72]

Block copolymer sample	Thickness/ ellipsometry	Thickness/SPS	Contact angle
PS-b-PI [a]	6 ± 2 nm (Au) 8 ± 2 nm (Si)	5.5 ± 0.1 nm (Au)	82°
PBd-b-PS [*,b]	8 ± 2 nm (Au)	12.0 ± 0.1 nm (Au)	80°

Using benzene/THF as solvent. All polymerizations were done at room temperature, 24–25 °C with 7 d for the reaction of each monomer block.
[a] PS-PI copolymer: 3.6 g styrene, 5.4 g isoprene, 2 mL THF, 5×10^{-3} moles s-BuLi.
[b] PBd-PS copolymer: 8.2 g styrene, 3.6 g butadiene, 2 mL THF, 5×10^{-3} moles s-BuLi.
[*] The PBd block was attached first to the surface followed by PS.

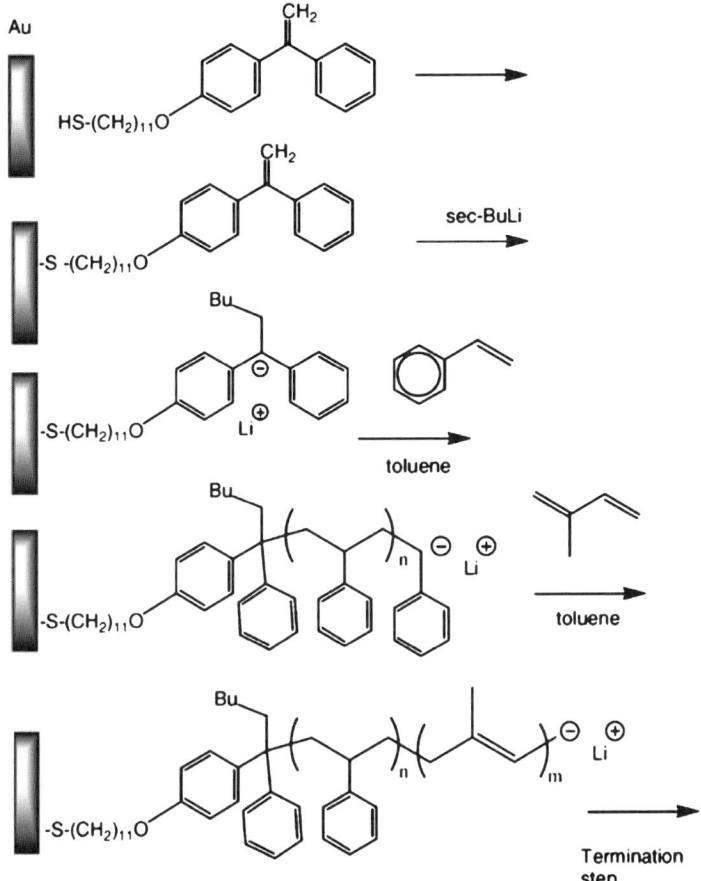

Fig. 7 Schematic diagram of block copolymer brush formation on a Au surface [72]

of the first monomer followed by polymerization of the second monomer. FTIR, PM–IRRAS, and XPS analysis confirmed the presence of **PS** and **PBd** blocks polymerized sequentially on the Au surface. In particular, to verify the presence of the first **PBd** block on the copolymer, we performed bromination experiments on the copolymer brush [81]. The XPS spectra indicated the presence of the new Br peaks [82]. The presence of a **PBd** block as shown by XPS and the second **PS** block by IR confirmed the polymerization of the second monomer (styrene) by the living chain ends of **PBd**. The sequence of polymerization can only allow grafting of *one polymer block at a time*. Also, the films have altogether a different morphology than **PS** brushes. Based on the FT-phase (amplitude) AFM images, these domains vary in their phase and depth but are not conclusive for the different block constitutions. Selective solvation and annealing may be necessary to observe other morphological features based on the surface energy of the blocks [83–87].

4
Cationic Surface-Initiated Polymerization

For cationic SIP, limitations when applied to surfaces are also evident. Like anionic polymerization, SIP on particles has been abundantly reported by several groups, most notably on work by Tsubokawa et al. The use of nanoparticles has also been widely reported. The same assumptions were made for cationic polymerization based on the grafting of electrophiles on surfaces and the reactivity of nucleophilic monomers for cationic polymerization.

4.1
Cationic SIP on Particles

The reported work on cationic SIP on particles includes the following:
(1) *Carbon blacks* have been reported to be capable of initiating the cationic polymerization of vinyl monomers such as vinyl ethers, indene, and acenaphthylene. The grafting sites of the polymer were based on carboxyl groups present on the surface [88]. The polymerization was inhibited by treatment of the carbon blacks with $NaHCO_3$, CH_2N_2, pyridine, and DMF. Also, the degree of conversion was found to be dependent on temperature and time of polymerization [89].
(2) *Cationic polymerization of N-vinylcarbazole and N-vinyl-2-pyrrolidone* initiated by carboxyl groups on carbon fibers has also been reported [90, 91]. The grafting ratio was measured at 39.7% (60 °C, 24 h) [92]. The effects of solvent and temperature on grafting were also investigated. The polymerization activation energy (E_a) was determined to be about 13.4 kcal/mol, and a similar loss in reactivity was found with treatment of $KHCO_3$, CH_2N_2, pyridine, and $HCONMe_2$, indicating the cationic nature of the reaction [93].
(3) *Polyesters have been grafted onto carbon black (whiskers) by cationic ring-opening polymerization of lactones* initiated by acylium perchlorate groups introduced onto the surface. Acylium perchlorate groups were introduced onto the surface of carbon whiskers, i.e., vapor-grown carbon fibers, by the reaction of surface acyl chloride groups with Ag perchlorate in $PhNO_2$. The benzylium perchlorate groups on the surface-initiated cationic ring-opening polymerization of cyclic ethers, such as THF and epichlorhydrin, and lactones, such as ε-caprolactone and δ-valerolactone, giving polyether- and polyester-grafted carbon black, respectively. Stable colloidal dispersions in good solvents were observed [94]. The rate of polymerization increased with increasing polymerization temperature, but the percentage of grafting onto the surface decreased [95]. The cationic graft polymerization of other vinyl monomers (e.g., styrene, indene, N-vinyl-2-pyrrolidone, and n-Bu vinyl ethers) onto carbon whiskers with acylium perchlorate surface groups was also investigated. The percentages of grafting were: polystyrene 42.5% and polyindene 100.3% [96].

(4) *Surface grafting of polymers onto ultrafine silica by cationic polymerization (initiated by benzylium perchlorate groups) has been demonstrated* [97]. The functionalization of silica with benzylium perchlorate was achieved by the reaction of Ag perchlorate with surface benzyl chloride groups (by the treatment of silica with 4-(chloromethyl)phenyltrimethoxysilane). The cationic SIP of styrene and cationic ring-opening SIP of ε-caprolactone, β-propiolactone were achieved. The percentage of grafting onto the silica surface decreased with increasing polymerization temperature, i.e., because chain transfer of the growing polymer cation accelerated with increasing temperature [98]. Cationic SIP by oxoaminium perchlorate groups introduced onto ultrafine silica surface was also demonstrated. The functional groups were successfully introduced by treatment of nitroxyl radicals on silica surface with $HClO_4$. Nitroxyl radicals were prepared by the reaction of 4-hydroxy-2,2,6,6-tetramethyl piperidinyloxy radical with acid anhydride groups on the surface. The cationic SIP of *iso*-Bu vinyl ether, N-vinylcarbazole, 2,3-dihydrofuran, and γ-butyrolactone were observed [99].

(5) *"Living like" cationic polymerization of isobutyl vinyl ether (IBVE) initiated by carboxyl groups was reported on carbon black surfaces/ethylaluminum dichloride systems in the presence of 1,4-dioxane.* The number average MW (Mn) of polyIBVE obtained was found to be directly proportional to monomer conversion in the cationic polymerization [100]. Living cationic polymerization of isobutyl vinyl ether by the carboxyl group of the carbon black/zinc chloride initiating system has also been investigated. Although the polymerization of IBVE was initiated by carboxyl groups on the surface, the rate of polymerization was small and the molecular weight distribution (MWD) of polyIBVE was very broad. The rate of the polymerization drastically increased, and 100% monomer conversion was achieved in a short time by the addition of $ZnCl_2$ [101].

(6) *2-Vinylthiophene (2-VT) has been cationically polymerized using chloroarylmethane derivatives grafted onto silica as reported by Spange et al.* The mass balance of the products (soluble fraction and hybrid particle fraction) was found to depend significantly on temperature and 2-VT/silica ratio. The transformation process of PVT toward conjugated polymers was also studied with UV-vis spectroscopy and ESR spectroscopy. The structure of PVT/silica and resulting hybrid materials was investigated by solid-state ^{13}C and ^1H CP MAS NMR-spectroscopy [102]. Cationic SIP of 2-vinylfuran with chlorotriphenylmethane as initiator has been used for the functionalization of these silica particles. The influence of monomer/initiator ratio and temperature on grafting efficiency, yield, and degree of grafting was investigated. Grafting efficiency was found to be a function of temperature and monomer/initator ratio, i.e., because crosslinking reactions between cationically active chains increase the amount of the immobilized polyvinylfuran fraction [103]. The cationic polymerization of electron-rich monomers such as vinyl ethers, vinyl furan, and cyclopentadiene on silica surfaces can be initiated by aryl methyl

~ 10 nm) of linear poly(N-propionylethylenimine) were of uniform thickness and were very stable [109].

(2) *The cationic polymerization of N-vinyl-2-pyrrolidone initiated electrochemically by anodic polarization on a Pt surface has been reported by Delhalle et al.* The electroinitiated polymerization of N-vinyl-2-pyrrolidone was investigated. The researchers observed that after electrolysis at controlled potentials, a thin, covering, and homogeneous film of poly(N-vinyl-2-pyrrolidone) appeared on the electrode. The insolubility of the polymer in typical organic solvents implied true chemical grafting onto the platinum surface. They reported that chain propagation occurred by a cationic mechanism initiated by a direct electron transfer leading to grafted cations on the surface, followed by the nucleophilic attack of neutral molecules. The MW of the polymer was estimated by GPC after mechanical removal of the film from the electrode. A MW distribution curve showed an average value of $Mn = 16\,000$ g/mol [110].

(3) *Photoinitiated interfacial cationic polymerization mechanisms and applications have been reported by Wilson et al.* Details of this interesting photomechanism were reported as follows. The UV-initiated interfacial cationic polymerization of 2-methoxypropene on surfaces of polystyrene and poly(p-methoxystyrene) films containing Ph_3SSbF_6 or Ph_3SAsF_6 was reported to proceed via a surface-phase add-layer mechanism in which the monomer diffused into the host polymer, where initiation occurred. This was followed by migration of the propagating chain ends to the surface where polymerization and rapid film growth occurred. The polymerization was observed to stop at long reaction times due to the trapping of chain ends inside the growing film. The resultant surface of the polymer films was hydrophilized by the interfacial polymerization process [111]. A highly nonlinear propagation (polymerization kinetics) was observed. The effect of reaction variables on the kinetics was discussed and a model was proposed [112]. Another surface modification was described consisting of doping the polymeric sample with a small amount of a soluble cationic photoinitiator, exposing the sample to UV irradiation to create a strongly acidic surface due to radiolysis of the photoinitiator and exposing this modified polymer to an appropriate monomer either in solution or in the gas phase. The photogenerated acid was observed to initiate cationic polymerization of the monomer, which polymerized on the surface to form a deposited film. This method was used to modify poly(p-methoxystyrene) and polystyrene with 2-methoxypropene, p-methoxystyrene, or Me vinyl ether using triarylsulfonium hexafluoroantimonate or -arsenate as initiators. In all cases, again, highly nonlinear polymerization kinetics was observed that included an induction period, rapid film growth, and a plateau in growth rate where polymerization stopped despite a constant monomer feed. The induction period was found to be related to the deactivation of catalyst by water and to a depletion of catalyst in the surface of the host polymer. The initiation occurred at some depth into the

host polymer, and the propagating chain ended its migration through the surface where they encountered available monomer, and rapid film growth occurred. The reaction slowed and eventually stopped because the chain ends were trapped inside the growing film [113].

(4) *The synthesis and characterization of polyisobutylene brushes on silicate substrates via carbocationic polymerization were detailed by Faust et al.* The synthesis of polyisobutylene (PIB) brushes on planar silicate substrates via living cationic polymerization was described by two approaches: "grafting from" and "grafting onto." The "grafting from" approach, whereby an initiating moiety is attached onto the substrate surface to start polymerization of monomers, yielded surface-bound polymer brushes with more uniform surface and higher grafting d compared to the "grafting to" technique. The effect of the MW of PIB and varying reaction time on the grafted thickness was also studied. For PIB brushes formed by "grafting from" the film thickness increased with increasing mol. wt. of PIB. In contrast, PIB brushes produced by "grafting onto" gave constant values with increasing MW of PIB due to steric repulsion [106, 107].

(5) *Lastly, Polystyrene (PS) brushes on silicate substrates were grafted via carbocationic polymerization from self-assembled monolayer (SAM) initiators as reported by Brittain et al.* The carbocationic initiators, 2-(4-(11-triethoxysilylundecyl))phenyl-2-methoxypropane and 2-(4-trichlorosilyl-phenyl)-2-methoxy-d_3-propane, and their corresponding SAMs were prepared on various substrates. The monolayers were characterized by FTIR–ATR, contact angles, and X-ray reflectometry. The growth of the PS brushes from

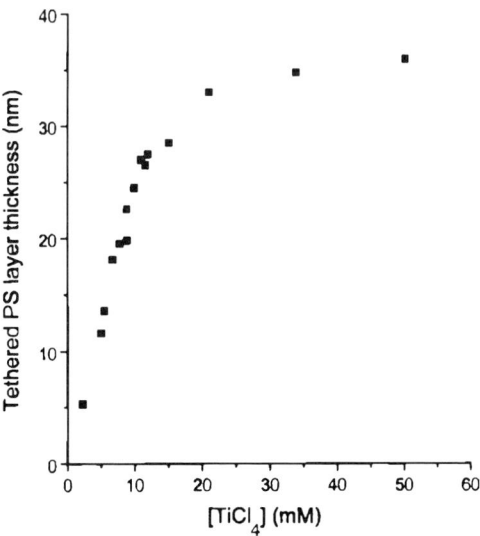

Fig. 10 Variation in thickness of polystyrene brush with [TiCl$_4$] [114]

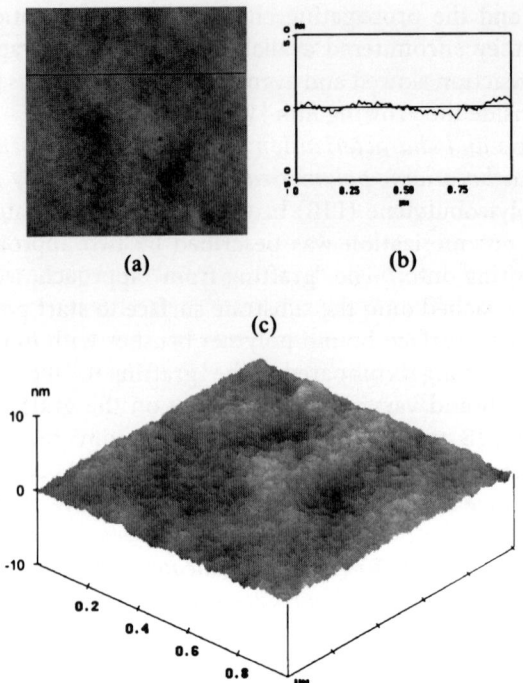

Fig. 11 Atomic force microscopy of a **a** 30-nm-thick grafted PS film, **b** depth profile along indicated *line*, and **c** 3D image [114]

these SAMs was observed to occur through the carbocationic polymerization mechanism. The parameters and factors that influence PS thickness including solvent polarity, additive, and TiCl$_4$ concentration were consistent to that of a typical cationic polymerization (Fig. 10). The sequential carbocationic polymerization on the same silicate substrate resulted in thicker PS films (addition of more monomers). FTIR–ATR measurements using a deuterated initiator indicated that the initiator efficiency is low, and the second carbocationic polymerization on the same sample involved initiation from both PS chain ends and unconsumed surface-immobilized initiators. However, AFM revealed a uniform and smooth PS brush surface with a roughness value of 0.3 nm (rms) for a 30-nm-thick PS layer (Fig. 11) [114].

5
Conclusions

In this article, the use of anionic and cationic surface-initiated methods to form homopolymer and block copolymer brushes grafted from flat and nanoparticle surfaces has been reviewed. The future challenge in understanding these methods is to differentiate the mechanisms from other addition-type polymerizations from surfaces where control of MW and polydispersity is important. Activation of the grafted initiator, control of polymerization conditions, and removal of excess activators are the common denominators with these methods. Some of the problems and potential of these techniques have been discussed. The SIP protocol requires surface initiators that can be grafted by SAM techniques. It is clear that the unique properties of these addition polymerization methods are based primarily on the formation of charged reactive propagating centers and controlled termination. The challenges in developing such techniques involves preserving the reactivity of the charged species, where the monomer, solvent quality, and lack of terminating species allow for grafting to surfaces and the formation of homopolymer and block copolymers. From what has been observed in the various systems reviewed, the mechanisms do not necessarily follow their bulk and solution counterparts. A focus was placed on the use of 1,1-diphenylethylene (DPE) coinitiators functionalized with alkylsilane or alkylthiol and grafted onto planar and particle surfaces using SAM techniques. Other anion initiators and initiator/monomer systems have shown promise in particular with SIP on particles and nanoparticles. For the cationic (carbocationic) SIP methods, most of the reported work is based on tethering Lewis acids to surfaces. Like anionic SIP, the polymerization involves a sequence of monomer introduction, reaction, and termination where homopolymer or block copolymer tethered polymer brushes can be obtained. In general, the grafted polymer chains have been investigated in situ using surface-sensitive spectroscopic and microscopic techniques. In order to obtain MW, polydispersity, and chemical structure data, e.g., by NMR, GPC, IR, etc., the polymer chains need to be removed from the substrate/particle surface. In this respect, SIP on particles is able to provide adequate polymer samples for ex situ analysis.

Acknowledgements I would like to acknowledge my collaborators in this research area: QingYe Zhou, Mi-Kyoung Park, Shuangxi Wang, Jimmy Mays, George Sakellariou, Stergios Pispas, and Nikos Hadjichristidis. I would like to acknowledge the Advincula research group and especially Derek Patton for help in preparing updated references. I would also like to acknowledge technical support from Molecular Imaging (Agilent technologies), Optrel GmBH, and Maxtec Inc.

References

1. Advincula R, Ruehe J, Brittain W, Caster K (eds) (2004) Polymer Brushes. Wiley, Weinheim, p 483
2. Advincula R (2003) J Dispersion Sci Tech 24:361
3. Advincula R (2004) Polymer Brushes. In: Kroschwitz J (ed) Encyclopedia of Polymer Science and Technology. Wiley, New York

4. Halperin A, Tirrell M, Lodge TP (1992) Adv Polym Sci 100:31
5. Tadros T (1982) The Effect of Polymers on Dispersion Properties. Academic, London
6. Krishnamoorti R, Vaia R (2002) Polymer Nanocomposites. ACS Symposium Series 804. Oxford University Press, Cary, NC
7. Prücker O, Rühe J (1998) Langmuir 14:6893
8. Zhao B, Brittain W (2000) Prog Polym Sci 25:677
9. Prücker O, Rühe J (1998) Macromolecules 31:592
10. Jordan R, Ulman A (1998) J Am Chem Soc 120:243
11. Weck M, Jackiw J, Rossi R, Weiss P, Grubbs R (1999) J Am Chem Soc 121:4088
12. Ejaz M, Yamamoto S, Ohno K, Tsujii Y, Fukuda T (1998) Macromolecules 31:5934
13. Kong X, Kawai T, Abe J, Iyoda T (2001) Macromolecules 34:1837
14. Yamamoto S, Tsujii Y, Fukuda T (2000) Macromolecules 33:5995
15. von Werne T, Patten T (2001) J Am Chem Soc 123:7497
16. Jones D, Brown A, Huck W (2002) Langmuir 18:1265
17. Husseman M, Malmstrom E, McNamara M, Mate M, Mecereyes D, Genoit G, Hedrick J, Mansky P, Huang E, Russell T, Hawker C (1999) Macromolecules 32:1424
18. Zhou Q, Nakamura Y, Inaoka S, Park M, Wang Y, Mays J, Advincula R (2000) Poly Math Sci Eng Preprint (Am Chem Soc) 82:291
19. Baum M, Brittain WJ (2002) Macromolecules 35:610
20. Sedjo RA, Mirous BK, Brittain WJ (2000) Macromolecules 33:1492
21. Odian G (2004) Principles of Polymerization, 4th edn. Wiley, New York
22. Pitsikalis M, Pispas S, Mays J, Hadjichristidis N (1998) Adv Polym Sci 135:1
23. Zhou Q, Wang S, Fan X, Pispas S, Sakellariou G, Hadjichristides N, Mays J, Advincula R (2001) Polym Preprints 42:59
24. Quirk R, Mathers R (2001) Polym Bull 6:471
25. Ingall M, Honeyman C, Mercure J, Bianconi P, Kunz R (1999) J Am Chem Soc 121:3607
26. Jordan R, Ulman A, Kang J, Rafailovich M, Sokolov J (1999) J Am Chem Soc 121:1016
27. Zhou Q, Fan X, Xia C, Mays J, Advincula R (2001) Chem Mater 13:2465
28. Zhou Q, Wang S, Fan X, Advincula R, Mays J (2002) Langmuir 18:3324
29. Fan X, Zhou Q, Xia C, Cristofoli W, Mays J, Advincula R (2002) Langmuir 18:4511
30. Hsieh H (1965) J Polym Sci A3:163
31. Fetters LJ, Morton M (1974) Macromolecules 7:552
32. Bhattacharyya DN, Lee CL, Smid J, Szwarc M (1965) J Phys Chem 69:612
33. Wakefield BJ (1974) The Chemistry of Organolithium Compounds. Pergamon, Oxford
34. Papirer E, Nguyen VT (1972) J Polym Sci Polym Lett Ed 10(3):167
35. Horn J, Hoene R, Hamann K (1975) Macromol Chem Suppl 1:329
36. Donnet JB, Papirer E (1975) Colloques Internationaux du Centre National de la Recherche Scientifique 231:117
37. Tsubokawa N, Yoshihara T, Sone Y (1991) Colloid Polym Sci 269:324
38. Tsubokawa N, Funaki A, Hada Y, Sone Y (1982) J Polym Sci Polym Chem Ed 20(12):3297
39. Tsubokawa N, Funaki A, Sone Y (1983) J Appl Polym Sci 28(7):2381
40. Tsubokawa N, Kobayashi K, Sone Y (1987) Polym J 19(10):1147
41. Tsubokawa N, Hamada H, Sone Y (1989) Polym-Plast Technol Eng 28(2):201
42. Tsubokawa N, Kogure A, Sone Y (1990) J Polym Sci Part A Polym Chem 28(7):1923–1933
43. Tsubokawa N, Hamada H, Fujiki K (1994) Polymer 35(5):1084
44. Tsubokawa N, Hamada H, Sone Y (1990) J Macromol Sci Chem A27(6):779

45. Tsubokawa N, Yoshihara T, Sone Y (1991) Colloid Polym Sci 269(4):324
46. Tsubokawa N, Yoshihara T, Sone Y (1992) J Polym Sci Part A Polym Chem 30(4):561
47. Raghavendran VK, Drzal LT (2002) Composite Interfaces 9(1):1
48. Schomaker E, Zwarteveen AJ, Challa G, Capka M (1988) Polym Commun 29:158
49. Husemann M, Morrison M, Benoit D, Frommer J, Mate M, Hinsberg W, Hedrick J, Hawker C (2000) J Am Chem Soc 122:1844
50. Milner ST (1991) Science 252:905
51. Milner ST, Witten TA, Cates ME (1988) Macromolecules 21:2610
52. Minko S, Gafijchuk G, Sidorenko A, Voronov S (1999) Macromolecules 32:4525
53. Alexander S (1977) J Phys 38:977
54. Rockford L, Mochrie S, Russell T (2001) Macromolecules 34:1487
55. Russell T, Thurn-Albrecht T, Tuominen M, Huang E, Hawker C (2000) Macromolecular Symp 159:77
56. Rockford L, Liu Y, Mansky P, Russell T, Yoon M, Mochrie S (1999) J Phys Rev Lett 82:2602
57. Zhulina EB, Balazs AC (1996) Macromolecules 29:6338
58. Fasolka M, Banerjee P, Mayes A, Pickett G, Balazs A (2000) Macromolecules 33:5702
59. Pereira G, Williams D (1999) Macromolecules 32:758
60. Ulman A (1991) An Introduction to Ultrathin Organic Films: From Langmuir-Blodgett to Self-Assembled Monolayers. Academic, Boston
61. Oosterling M, Sein A, Schouten A (1992) Polymer 33(20):4394
62. Morton M, Fetters L (1975) J Rubber Chem Technol 48:359
63. Quirk RP, Mathers RT (2001) Polym Mater Sci Eng 84:873
64. Quirk RP, Mathers RT (2001) Polym Mater Sci Eng 85:198
65. Zhou Q, Nakamura Y, Inaoka S, Park M, Wang Y, Mays J, Advincula R (2000) Polym Mater Sci Eng 82:290
66. Advincula R, Zhou Q, Mays J (2001) Polym Mater Sci Eng 84:875
67. Zhou Q, Fan X, Xia C, Mays J, Advincula R (2001) Polym Mater Sci Eng 84:835
68. Quirk RP, Yoo T, Lee Y, Kim J, Lee B (2000) Adv Polym Sci 153:67
69. Zhou Q, Nakamura Y, Inaoka S, Park M, Wang Y, Mays J, Advincula R (2002) In: Polymer Nanocomposites, Krishnamoorti R, Vaia R (eds) ACS Symposium Series 804. Oxford University Press, Cary, NC
70. Glasse MD (1983) Prog Polym Sci 9:133
71. Quirk R, Mathers R, Cregger T, Foster M (2002) Macromolecules 35:9964
72. Advincula R, Zhou Q, Park MK, Wang S, Mays J, Sakellariou G, Pispas S, Hadjichristidis N (2002) Langmuir 18:8672
73. Förster S, Krämer E (1999) Macromolecules 32:2783
74. Hadjichristidis N, Iatrou H, Pispas S, Pitsikalis M (2000) J Polym Sci A Polym Chem 38:3211
75. Wirth M, Fairbank R, Fatunmbi H (1997) Science 275:44
76. Fadeev AY, McCarthy TJ (1999) Langmuir 15:3759
77. Prucker O, Ruhe J (1998) Langmuir 14:6893
78. Szwarc M (1956) Nature 178:1168
79. Wittmer J, Cates M, Jhoner A, Turner M (1996) Europhys Lett 33:397
80. Knoll W (1998) Annu Rev Phys Chem 49:569
81. Buzdugan E, Ghioca P, Badea E, Serban S, Stribeck N (1997) Eur Polym J 33:1713
82. Galuska A (1999) Surf Interface Anal 27:889
83. Iwata H, Hirata I, Ikada Y (1997) Langmuir 13:3063
84. Zhao B, Brittain WJ (2000) Macromolecules 33:342
85. Zhao B, Brittain WJ (2000) Macromolecules 33:8813

86. Zhao B, Brittain WJ, Zhou W, Cheng SZD (2000) J Am Chem Soc 122:2407
87. Sidorenko A, Minko S, Schenk-Meuser K, Duschner H, Stamm M (1999) Langmuir 15:8349
88. Tsubokawa N, Takeda N, Iwasa T (1981) Polym J 13(12):1093
89. Tsubokawa N (1980) J Polym Sci Polym Lett Ed 18(6):461
90. Tsubokawa N, Maruyama H, Sone Y (1988) J Macromol Sci Chem A25(2):171
91. Tsubokawa N, Yoshihara T (1991) Polym J 23(3):177
92. Tsubokawa N, Maruyama H, Sone Y (1986) Polym Bull 15(3):209
93. Tsubokawa N, Takeda N, Kanamaru A (1980) J Polym Sci Polym Lett Ed 18(9):625
94. Tsubokawa N, Handa S (1993) J Macromol Sci Pure Appl Chem A30(4):277
95. Tsubokawa N, Yoshihara T (1993) Polym Bull 30(4):421
96. Tsubokawa N, Yoshihara T (1993) J Macromol Sci Pure Appl Chem A30(8):517
97. Tsubokawa N, Kogure A (1993) Polym J 25(1):83
98. Tsubokawa N, Saitoh K, Shirai Y (1995) Polym Bull 35(4):399
99. Tsubokawa N, Kimoto T, Endo T (1994) Polym Bull 33(2):187
100. Tsubokawa N, Oyanagi K, Yoshikawa S (2000) J Macromol Sci Pure Appl Chem A37(6):529–548
101. Yoshikawa S, Nishizaka R, Oyanagi K, Tsubokawa N (1995) J Polym Sci Part A Polym Chem 33(13):2251
102. Hoehne S, Seifert A, Friedrich M, Holze R, Spange S (2004) Macromol Chem Phys 205(12):1667
103. Hoehne S, Spange S (2003) Macromol Chem Phys 204(5/6):841
104. Spange S, Eismann U, Hoehne S, Langhammer E (1998) Macromolecular Symposia (6th Dresden Polymer Discussion Surface Modification, 1997) 126:223
105. Jordan R, West N, Ulman A, Chou YM, Nuyken O (2001) Macromolecules 34(6):1606
106. Kim IJ, Faust RJ (2003) Macromol Sci Pure A40(10):991
107. Kim IJ, Angelopoulos A, Faust R (2001) Polym Preprints 42(2):481
108. Wang WP, Pan CY (2004) Polymer 45(12):3987
109. Jordan R, Ulman A (1998) J Am Chem Soc 120(2):243
110. Leonard-Stibbe E, Lecayon G, Deniau G, Viel P, Defranceschi M, Legeay G, Delhalle J (1994) J Polym Sci Part A Polym Chem 32(8):1551
111. MacDonald S, Hult A, Allen R, Wilson CG (1985) Proceedings of the International Conference on Organic Coatings, Science and Technology, New Paltz, NY, p 203
112. MacDonald S, Hult A, Wilson CG (1985) Macromolecules 18(10):1804
113. MacDonald S, Hult A, Allen R, Wilson CG (1985) Polym Mater Sci Eng 52:339
114. Zhao B, Brittain WJ (2000) Macromolecules 33(2):342

Metathesis Polymerization To and From Surfaces

Michael R. Buchmeiser[1,2]

[1]Leibniz Institut für Oberflächenmodifizierung e.V., Permoserstr. 15, 04318 Leipzig, Germany
michael.buchmeiser@iom-leipzig.de

[2]Technische Chemie der Polymere, Universität Leipzig, Linnéstraße 2, 04103 Leipzig, Germany
michael.buchmeiser@iom-leipzig.de

1	Introduction to Ring-Opening Metathesis Polymerization (ROMP) and 1-Alkyne Polymerization	138
2	Initiators Suitable for Surface-Initiated ROMP	140
2.1	Molybdenum-Based Initiators (Schrock Catalysts)	140
2.2	Ruthenium-Based Initiators (Grubbs, Grubbs-Herrmann, Grubbs-Hoveyda Catalysts)	141
3	Supports	142
3.1	Inorganic Surfaces	142
3.1.1	Gold Surfaces	142
3.1.2	Silicon- and Silica-Based Surfaces	143
3.1.3	Other Inorganic Surfaces	154
3.2	Organic Surfaces	155
3.2.1	Merrifield-Type Resins	155
3.2.2	Monolithic Supports	156
3.2.3	Other Organic Surfaces	167
4	Conclusion/Outlook	167
	References	168

Abstract Polymerization to and from surfaces using metathesis-based techniques – ring-opening metathesis polymerization (ROMP) and 1-alkyne polymerization – is reviewed. The application of both "*grafting from*" and "*grafting to*" techniques along with defined catalytic systems suitable for these purposes is discussed. Special consideration will be given to the specific properties of silica, silicon and organic supports and the resulting requirements for the polymerization catalysts used. The surface-modified materials obtained will be discussed in view of their applications, which mainly fall into the areas of separation science, catalysis and molecular electronics.

Keywords 1-Alkyne polymerization · Grafting · Ring-opening metathesis polymerization (ROMP) · Surfaces

Abbreviations
ATRP   atom-transfer radical polymerization
β-CD β-cyclodextrin

CMP	contact metathesis polymerization
CSP	chiral stationary phase
CTA	chain-transfer agent
DCPD	dicyclopentadiene
DEDAM	diethyl diallylmalonate
DMAP	dimethylaminopyridine
DMF	dimethylformamide
DMN-H6	1,4,4a,5,8,8a-hexahydro-1,4,5,8-*exo-endo*-dimethanonaphthalene
DNS	dansyl
EVE	ethyl vinyl ether
Fmoc	fluorenylmethoxycarbonyl
ICP-OES	inductively coupled plasma-optical emission spectroscopy
ISEC	inverse size-exclusion chromatography
NBDE	norbornadiene
NHC	*N*-heterocyclic carbene
PDI	polydispersity, polydispersity index
PS-DVB	poly(styrene-*co*-divinylbenzene)
RCM	ring-closing metathesis
SAM	self-assembled monolayer
ROMP	ring-opening metathesis polymerization
SPE	solid-phase extraction
THF	tetrahydrofuran
TOF	turnover frequency
TON	turnover number

1
Introduction to Ring-Opening Metathesis Polymerization (ROMP) and 1-Alkyne Polymerization

The polymerization of cyclic, strained olefins by transition metal alkylidenes of general formula $L_nM = CRR'$ (L = ligand, R, R' = H, alkyl, aryl) yields polymers formed via ring-opening that contain unsaturated double bonds within each repetitive unit. Since the mechanism is based on repetitive metathesis steps, this polymerization reaction is known as "ring-opening metathesis polymerization" (ROMP) (Scheme 1).

For olefins, cyclic, or better bi- or tricyclic ring structures with large ring strain (norborn-2-enes or norbornadienes for instance) are required. Alternatively, 1-alkynes can be used. In this case, the term "1-alkyne polymerization" applies. This reaction proceeds via α- or β-insertion of the alkyne into the metal–carbon double bond (Scheme 1). Both insertion mechanisms lead to a conjugated polymer. With a few exceptions [1–3], polymerizations based on α-insertion are the preferred ones, since they offer better control over molecular weights due to favorable values of k_i/k_p (k_i, k_p = rate constants of initiation and propagation, respectively).

A number of catalysts or (better) initiators that can accomplish ROMP have been developed over the last 25 years [4]. Nevertheless, only a few of

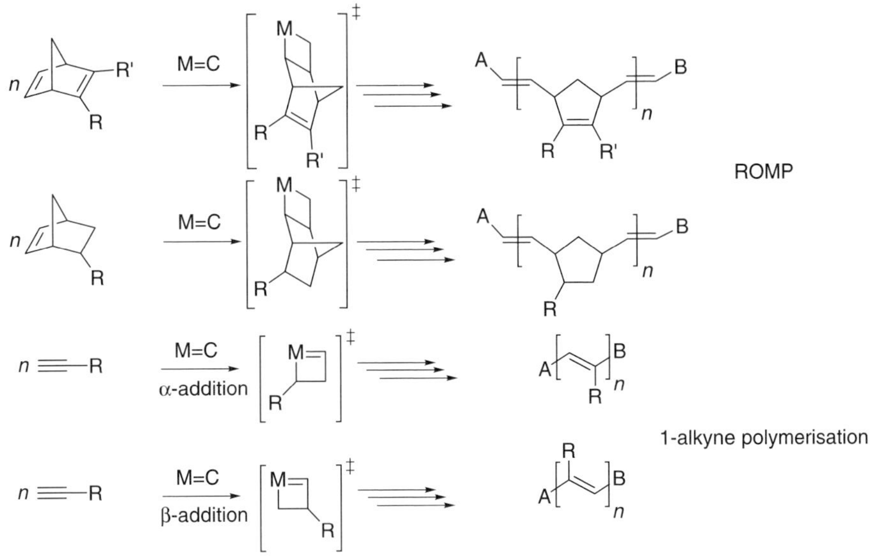

Scheme 1 ROMP of a 2,3-disubstituted norbornadiene, a 2-substituted norborn-5-ene and polymerization of a 1-alkyne via α- and β-addition, respectively. A and B are initiator- and termination-derived endgroups, respectively

them can be used to generate truly living systems. Among these, the two most prominent systems are based on either molybdenum or ruthenium and are referred to as Schrock [5] and Grubbs [6] catalysts, respectively. Numerous variants exist that allow the generation of class VI living ROMP systems in most cases, where chain transfer as well as uncontrolled chain-terminating reactions are totally absent [7]. Until recently, 1-alkyne polymerization had been the sole domain of Schrock catalysts (vide infra) [8]. By carefully tuning of both the steric and the electronic properties of the catalysts, they can be used to generate polymerization systems where k_p is comparable to k_i, resulting in polymerizations characterized by the complete and instantaneous consumption of the initiator. When an initiator fulfills all of these criteria, polymers with defined molecular weights and low polydispersities (PDI, typically < 1.15) are obtained. In addition, stoichiometric design and block copolymer construction are possible. Finally, the formation of a specific backbone structure – the relative orientation of one monomer unit to another – can be predetermined by choosing a particular initiator. Nevertheless, in order to rely on these potential advantages, a careful investigation of the polymerization system is necessary for each monomer used.

Even more importantly, in combination with well-defined initiators ROMP can be used to polymerize *functional* monomers. Comprehensive descriptions of such processes have been given in numerous reviews, and these far exceed the scope of this contribution [4, 6, 9–13]. Nevertheless, the most important features relevant to ROMP in combination with surfaces will be briefly summarized in the following.

2
Initiators Suitable for Surface-Initiated ROMP

2.1
Molybdenum-Based Initiators (Schrock Catalysts)

Molybdenum-based systems of general formula $Mo(N-Ar')(CHCMe_2R)(OR')_2$ ($Ar' = 2,6\text{-}Me_2-C_6H_3$, $2,6\text{-}iPr_2-C_6H_3$; $R = Me$, Ph; $R' = CMe_3$, $CMe(CF_3)_2$), usually referred to as Schrock initiators, represent highly active initiators, particularly when used in combination with electronegatively substituted alkoxides (Fig. 1).

Due to their high reactivity and the pronounced oxophilic character of molybdenum, they are not capable of polymerizing monomers containing protic hydrogens such as alcohols, carboxylic acids, thiols or aldehydes and ketones. Nonetheless, a broad range of functional monomers based on 2-substituted and 2,3-disubstituted norborn-5-enes and norbornadienes bearing anhydrides, esters, amides, pyridines, and so on can be polymerized. Particular advantages of these initiators are their tunable (and generally high) activity, the ease of end group functionalization via a Wittig-type reaction

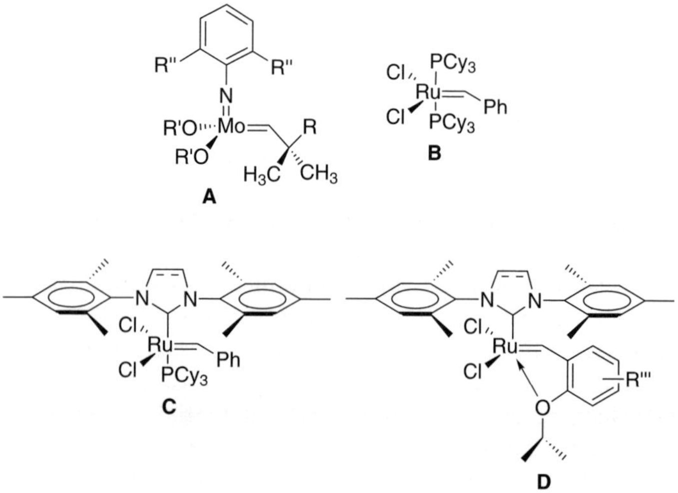

Fig. 1 Well-defined metathesis initiators. Schrock catalyst (**A**, $R = Me$, Ph, $R' = CMe_3$, $CMe(CF_3)_2$), Grubbs-catalyst (**B**), Grubbs-Herrmann catalyst (**C**), Grubbs-Hoveyda catalyst (**D**, $R''' = H$, NO_2)

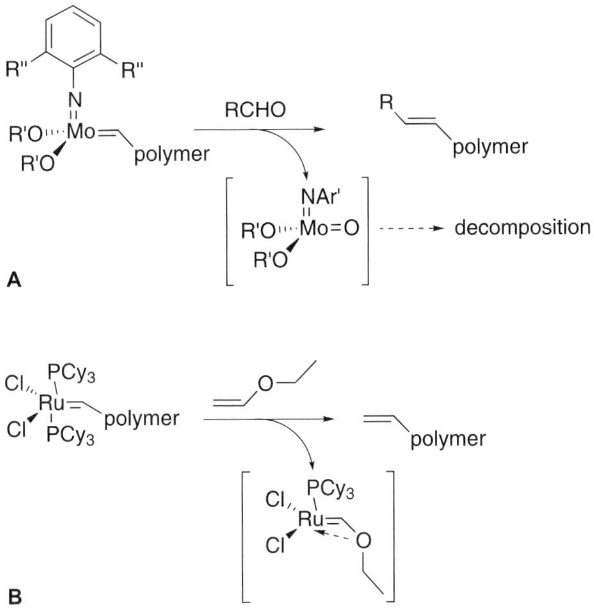

Scheme 2 Termination of living polymers. **A** Wittig-type reaction for Mo-based initiators, **B** metathesis with ethyl vinyl ether for Ru-derived initiators

with aldehydes (Scheme 2), as well as their applicability to 1-alkyne polymerization. However, if used in ROMP or 1-alkyne polymerization to or from surfaces, particular attention must be devoted to preventing any contact with acidic sites such as silanol groups that are located on the surface.

2.2
Ruthenium-Based Initiators
(Grubbs, Grubbs-Herrmann, Grubbs-Hoveyda Catalysts)

A more robust alternative is provided by the ruthenium-based systems developed by Grubbs et al. (Fig. 1). While strong donor ligands such as nitriles or pyridines inhibit any polymerization activity for $RuCl_2(PCy_3)_2(CHPh)$, the Grubbs-Herrmann [14–17] and Grubbs-Hoveyda [18–20] versions in particular are robust yet highly active polymerization initiators that sometimes rival the activity of molybdenum initiators [21]. These exhibit higher stability towards protic functionalities than molybdenum systems. In addition, ruthenium is far less oxophilic than molybdenum, which permits the polymerization of functionalized norborn-2-enes and 7-oxanorborn-2-enes, even in aqueous media [22–27]. Termination of these initiators is usually accomplished by adding excess ethyl vinyl ether (EVE) (Scheme 2).

3
Supports

3.1
Inorganic Surfaces

3.1.1
Gold Surfaces

The first surface modification using a "grafting-from" approach, as reported by Nguyen et al. They used 1-mercapto-10-(*exo*-5-norborn-2-enoxy)decane-modified gold nanoparticles for the $RuCl_2(PCy_3)_2(CHPh)$-initiated grafting of ferrocene-containing norborn-2-enes to produce redox-active polymer–nanoparticle hybrids (Scheme 3) [28, 29]. The concept was later extended to insulating surfaces such as silicon using a synthetic protocol similar to the one described by Grubbs et al. and our group (vide infra) [30].

A similar approach to modifying Au-surfaces was reported by Grubbs et al., who used the more rigid tether molecule 4-(4-(norborn-5-ene-2-ylmethylenoxy)phenylethynyl)tolane-4'-thiol, shown in Fig. 2. *N*-Methyl-7-oxanorborn-5-ene-5,6-dicarbimide and 2,3-bis(*tert*-butoxydimethylsilyloxymethylene)-norborn-5-ene were used in a "grafting-from" approach [31].

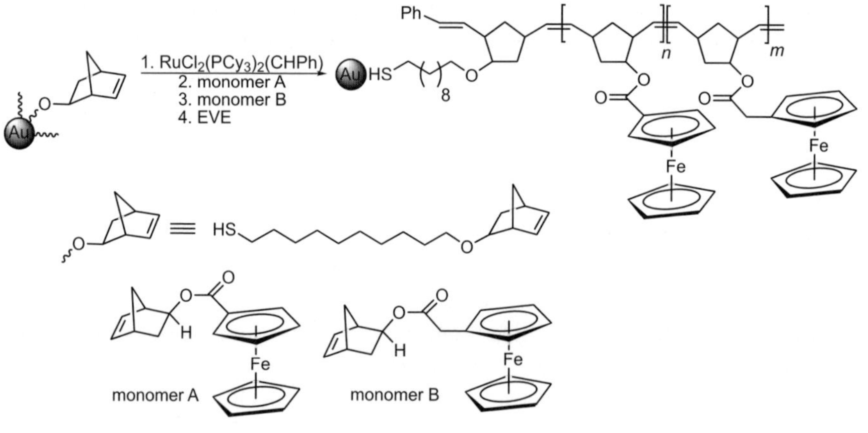

Scheme 3 Synthesis of Au-hybrid nanoparticles with electroactive copolymer shell structure

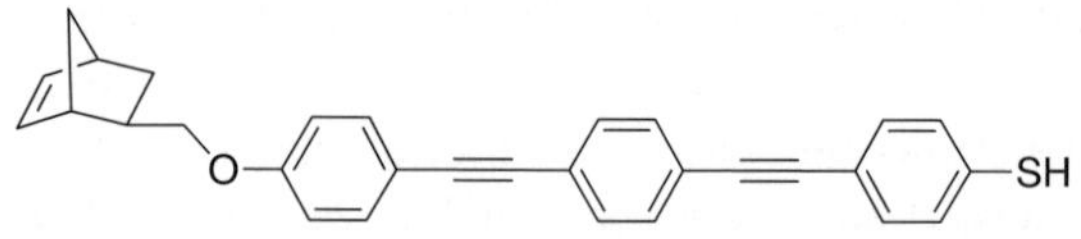

Fig. 2 4-(4-(Norborn-5-ene-2-ylmethylenoxy)phenylethynyl)tolane-4'-thiol used as anchor group in a "grafting-from" approach to the surface modification of Au surfaces

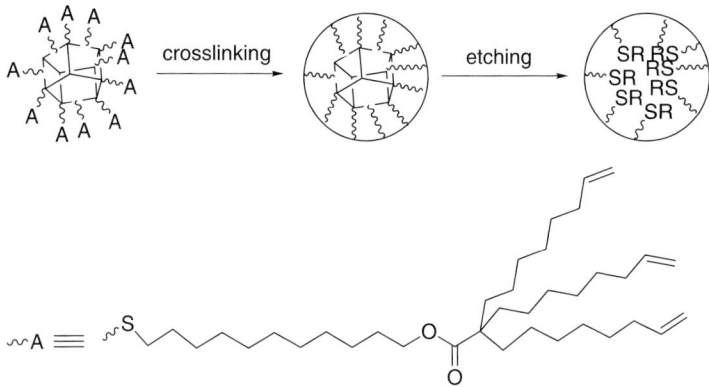

Fig. 3 Synthesis of nanometer-sized hollow polymer capsules from polymer-coated Au-particles

Similarly, gold particles were surface-modified with norborn-5-ene-2-ylmethanthiol and the surface-immobilized norborn-2-ene groups were subsequently used for norborn-2-ene polymerization in a "grafting-from" approach. The resulting material was used to construct a field effect transistor [32].

Reinhoudt et al. utilized a norborn-5-en-2-yloxydodecan-1-thiol-protected gold cluster. Crosslinking of the core was accomplished with $RuCl_2(PCy_3)_2$ (CHPh). Both intra- and interparticle crosslinking was observed [33]. The latter can be avoided by employing flat Au surfaces. Shultz et al. described the synthesis of nanometer-sized hollow polymer capsules from polymer-grafted gold particles. A metathesis route employing Grubbs' first generation initiator was applied, where the terminal alkene groups of the tripodal Au-immobilized ligand shown in Fig. 3 were crosslinked to form a three-dimensional polymer network. Etching of the gold particles with $KCN/K_2[Fe(CN)_3]/THF$ yielded the desired hollow capsules [34].

Finally, Sarkar et al. reported on the synthesis of Au nanoparticles carrying self-assembled monolayers (SAMs) with various functional groups including ferrocene and Fischer carbenes (Scheme 4) [35].

3.1.2
Silicon- and Silica-Based Surfaces

Many applications of surface modified materials (such as in molecular electronics, separation science or continuous flow catalysis) require the use of mechanically and pressure-stable carriers. Grubbs et al. and later Nuzzo et al. reported on the surface modification of Si(111). Conversion of surface Si – H into Si–allyl groups allowed them to pursue the "grafting-from" approach shown in Scheme 5 [36, 37]. The thickness of the polymer layer could be

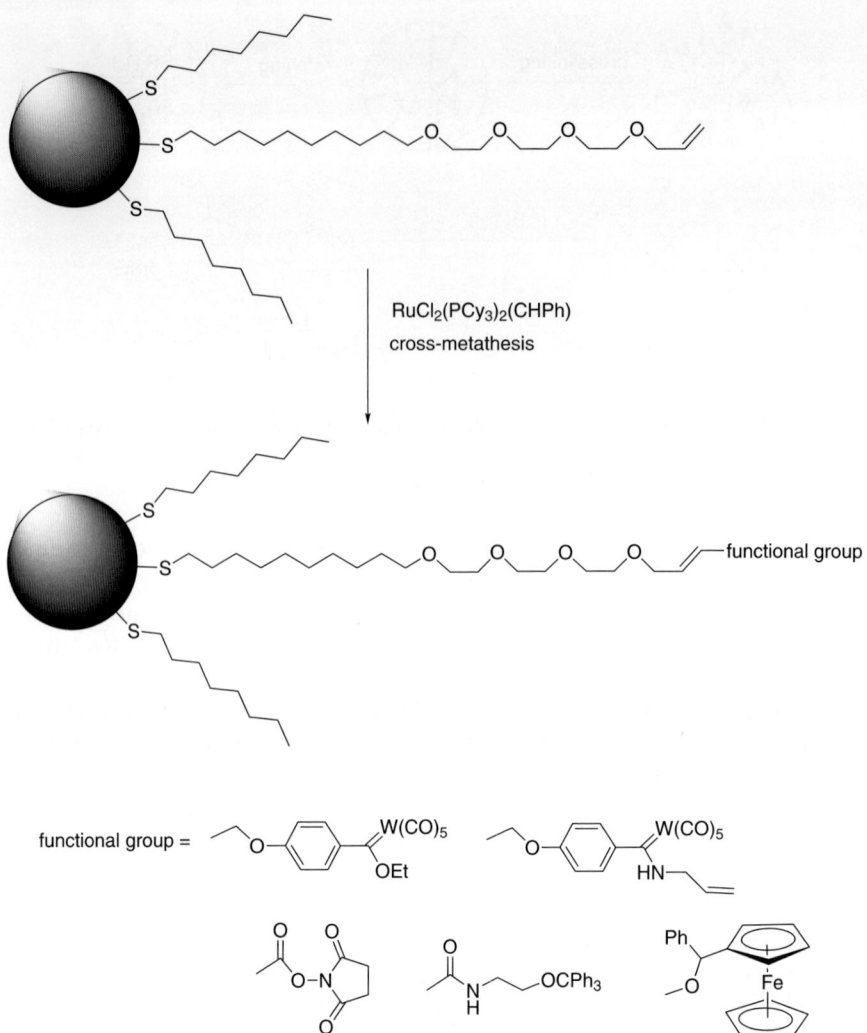

Scheme 4 Synthesis of Au nanoparticles bearing functional self-assembled monolayers (SAMs)

varied by up to 5500 nm by simply varying the monomer (norborn-2-ene) concentration.

Our group was the first to develop synthetic protocols for both "grafting-from" and "grafting-to" approaches to the modification of micrometer-sized inorganic (silica) particles (Scheme 6) [38–40].

Surface-immobilized norborn-5-ene-2-yl-groups were used as suitable anchoring groups for the preparation of ROMP-graft-copolymers. These can easily be introduced in the case of silica materials using trichloro-norborn-

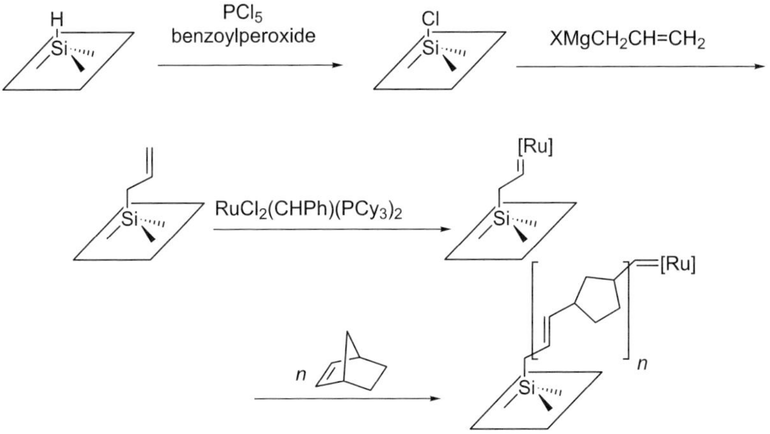

Scheme 5 Surface modification of Si using a "grafting-from" approach

5-ene-2-ylsilane, chlorodimethylnorborn-5-ene-2-ylsilane or trialkoxy-norborn-2-ene-5-ylsilanes. The former appears most favorable for gaining access to accurate surface analysis via elemental analysis, since all carbon found in norborn-5-ene-2-ylsilyl-derivatized silica can be clearly attributed to the surface-immobilized norborn-2-ene groups. In contrast, the use of trialkoxy-norborn-2-ene-5-ylsilanes results in the formation of additional surface-bound alkoxysilanes that impede accurate quantification of surface-bound norborn-2-ene groups via elemental analysis [41, 42]. Subsequent "endcapping" with a mixture of chlorotrimethylsilane and dichlorodimethylsilane followed by addition of absolute methanol leads to sufficient derivatization of a major part of the surface silanol groups (approximately 90%). For the "grafting-to" approach, the monomer was transformed into a living polymer via ROMP and subsequently attached to the support by reacting it with the surface norborn-2-ene groups. This approach required at least class IV living systems [7] and leads to the formation of tentacle-type stationary phases with the linear polymer chains attached to the support. Alternatively, the initiator can first be reacted with the support, making it heterogenized. Monomer is added consecutively and grafted onto the surface ("grafting-from" approach). While the first-generation Grubbs-type initiator $RuCl_2(PCy_3)_2(CHPh)$ could only be used for "grafting-from" experiments, Schrock-type initiators are applicable to both methods.

Using these methods, various monomers (such as N-(norborn-5-ene-2-carboxyl)-phenylalanine ethylester) were surface-grafted onto porous 5 μm silica. Using Nucleosil 300-5, 60 μmol of this monomer were immobilized on the surface using a "grafting-from" approach and $RuCl_2(PCy_3)_2(CH-p-F-C_6H_4)$ as initiator [38]. With $Mo(N-2,6-Me_2-C_6H_3)(CHCMe_2Ph)(OCMe(CF_3)_2)_2$, 40 μmol of this monomer could be grafted to the surface using either a "grafting-from" or a "grafting-to" approach. The resulting chiral station-

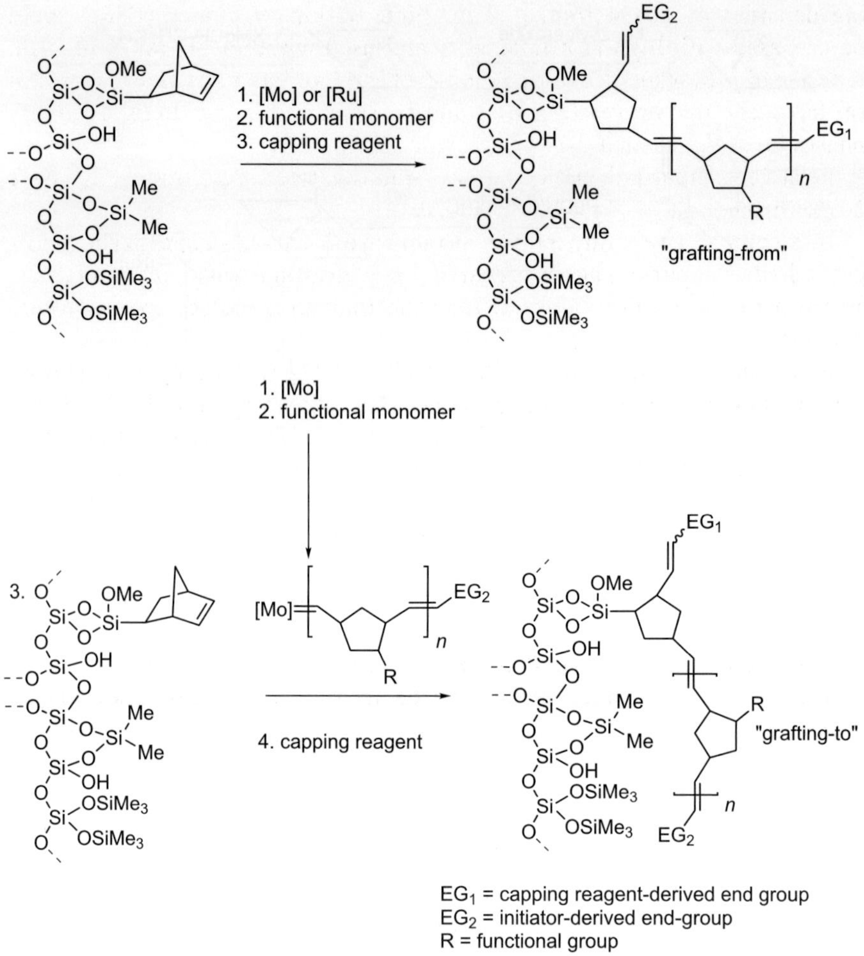

Scheme 6 Surface functionalization of silica via ROMP. "Grafting-from" approach (*top*), "grafting-to" approach (*bottom*)

ary phase was successfully used as an HPLC support in the separation of racemic dinitrobenzoyl-protected phenylalanine ethylester [38]. The broad applicability of this concept was demonstrated by immobilizing a series of β-cyclodextrin (β-CD) derivatives, 6-O-(norborn-2-ene-5-carboxyl)-β-CD, tetrakis(6-O-norborn-2-ene-5-carboxyl)-β-CD, 6-O-(7-oxanorborn-2-ene-5-carboxyl)-β-CD, 6-O-(6-(norborn-2-ene-5-carbonylaminohexoyl)-β-CD, 6-O-(norborn-2-ene-5-ylmethoxymethylsilyl)-β-CD, tris(6-O-norborn-2-ene-5-ylmethoxymethylsilyl)-β-CD, tetrakis(6-O-norborn-2-ene-5-ylmethoxymethylsilyl)-β-CD and hexakis(6-O-norborn-2-ene-5-ylmethoxymethylsilyl)-β-CD, on Nucleosil 300-5 [43]. A "grafting-from" approach using $RuCl_2(PCy_3)_2(CHPh)$ as initiator was used throughout, resulting in grafting densities of 11–34 µmol/g. The chiral stationary phases (CSPs) could be prepared with high reproducibility and used within a pH range of 2–10. A series of β-blockers, DNS-or Fmoc-protected amino acids and planar chiral ferrocene derivatives (Fig. 4) could be separated using these stationary phases [43, 44]. Selected data are provided in Table 1.

Relative standard deviations (σ_{n-1}) for the mean resolution (R_s) were 2–7% throughout.

In a comparative study, poly(7-oxanorborn-5-ene-2,3-dicarboxylic acid)-grafted silica supports, again prepared via a "grafting-from" approach, possessed better separation behavior than the analogous coated separation media [45].

Based on our studies on metallocenylalkynes [1–3, 46], poly(ethynylferricinium)-based anionic exchangers were prepared by applying the "grafting-to" concept described above [47]. Thus, metathesis polymerization of 4-ethynyl-1-(octamethylferrocenylethenyl)benzene using the Schrock-type catalyst Mo(N-2,6-Me$_2$ – C$_6$H$_3$)(CHCMe$_2$Ph)(OCMe(CF$_3$)$_2$)$_2$ and subsequent grafting of the living polymer onto a (norborn-5-ene-2-yl)-derivatized silica support resulted in the desired octamethylferrocene-grafted stationary phase (Scheme 7). Both porous (Nucleosil 300-5) and nonporous (Micra) silica was used.

Oxidation with iodine resulted in an octamethylferricinium-based anion-exchanger that was successfully used to separate oligonucleotides (dT_{12}–dT_{18}).

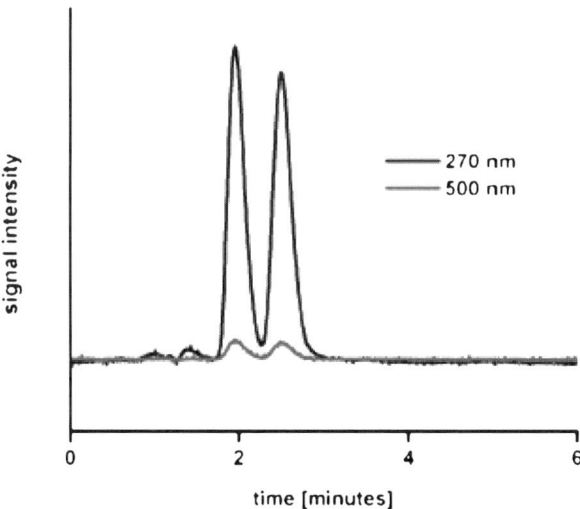

Fig. 4 Separation of *rac*-ferroceno[2,3a]inden-1-ones on a poly(tetrakis(*endo/exo*-6-O-norborn-2-ene-5-ylmethoxymethylsilyl)-β-CD)-grafted column. Conditions: $T = 21.5\,°C$, flow: 0.5 mL/min, acetonitrile-MeOH-acetic acid-triethylamine (90 : 10 : 0.15 : 0.45), UV-detection

Table 1 Separation of racemic compounds on 6-*O*-(norborn-2-ene-5-carboxyl)-β-CD-grafted Nucleosil 300-5

	k_L	k_D	α	R_S
DNS-Val [a]	2.03	5.07	2.50	4.75
DNS-Trp [a]	3.52	5.35	1.52	2.12
DNS-Thr [a]	0.47	1.44	3.08	2.38
DNS-Ser [a]	0.86	1.59	1.85	1.08
DNS-Phe [a]	3.10	5.02	1.62	2.50
DNS-Met [a]	1.72	3.30	1.92	2.71
DNB-Val [a]	4.32	5.50	1.27	1.22
DNB-Trp [a]	6.61	7.88	1.19	0.89
DNB-Phe [a]	10.21	6.84	1.49	2.36
Fmoc-Phe [a]	10.15	9.09	1.12	0.67
atenolol [b]	9.09	11.68	1.28	1.05
propranolol [b]	2.19	2.60	1.19	0.56
metoprolol [b]	2.57	3.27	1.27	0.85
proglumide [b]	2.08	3.54	1.70	2.57

Column dimensions: 150×2 mm; $T = 0$ °C; flow = 0.5 ml/min;
[a] 99.8/0.2/0.01/0.03 acetonitrile/MeOH/acetic acid/triethylamine;
[b] 98/2/0.2/0.2 acetonitrile/MeOH/acetic acid/triethylamine

As shown in preceding investigations, *N*,*N*-dipyrid-2-ylnorborn-2-en-5-ylcarbamide can be polymerized in a living manner using well-defined Schrock initiators [48]. Thus, a class VI living system [7] was accomplished with Mo(N-2,6-*i*-Pr$_2$ – C$_6$H$_3$)(CHCMe$_2$h)(CMe(CF$_3$)$_2$)$_2$. This monomer was grafted onto norborn-2-ene surface-functionalized silica, using a "grafting-from" approach to generate tentacles of poly-(*N*,*N*-dipyrid-2-ylnorborn-2-en-5-ylcarbamide) with a controlled degree of polymerization (DP), typically < 50 (Scheme 8) [49].

It is worth mentioning that the careful endcapping of silica with a mixture of ClSiMe$_3$ and Cl$_2$SiMe$_2$ eliminates any initiator deterioration potentially caused by the interaction with the silanol groups. In addition, complete reaction of the initiator with the support, as evidenced by the absence of any soluble polymer, was observed [38]. Palladium-loading of the supports was simply accomplished by reacting with H$_2$PdCl$_4$. A quantitative reaction was observed within a few hours, resulting in slightly yellow-colored supports. Values of 0.28 mmol and 0.08 mmol Pd/g, respectively, were achieved. Not surprisingly, RuCl$_2$(PCy$_3$)$_2$(CHPh) was not capable of polymerizing *N*,*N*-dipyrid-2-ylnorborn-2-en-5-ylcarbamide or its 7-oxa analog due to irreversible coordination of the ligand to the ruthenium core. The palladium-loaded silica was successfully used in various Heck reactions in-

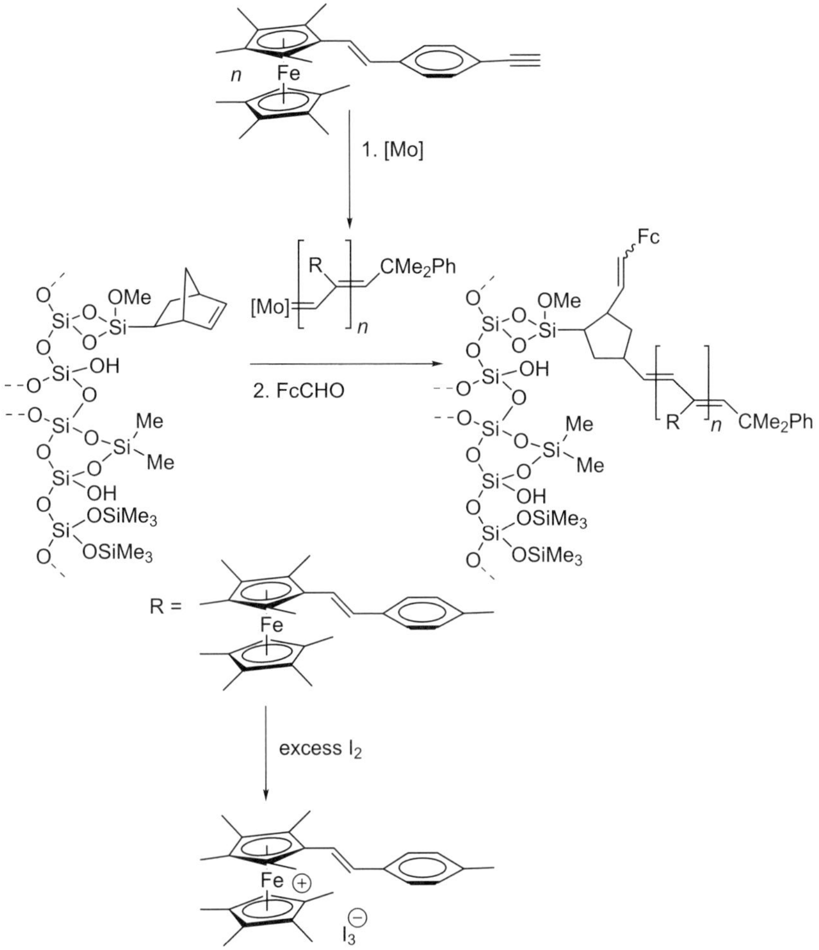

Scheme 7 Surface functionalization of silica via 1-alkyne polymerization using a "grafting-to" approach

cluding slurry reactions under standard as well as under microwave conditions. Removal of the support was simply accomplished by filtration. In particular, the use of microwaves lead to a drastic reduction in the reaction time, which is of particular interest for applications to high-throughput screening (HTS). Turnover frequencies (TOFs) were typically in the range of $0.1–0.3\,\mathrm{s}^{-1}$. Alternatively, palladium-loaded silica was packed into stainless steel columns which were subsequently loaded with Heck monomers and used as reaction columns in HTS machines. Alternatively, flow-through reactors were realized with surface-derivatized silica-packed stainless steel columns. With these columns, a constant conversion of iodobenzene with styrene (70–80%) was observed over several hours. TOFs were in the range of

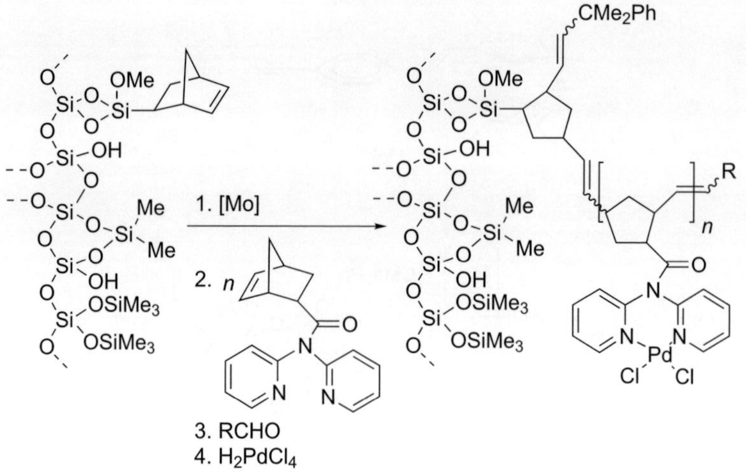

Scheme 8 Immobilization of norborn-5-ene-5-N,N-dipyrid-2-ylcarbamide on silica-60 using a "grafting-from" approach

0.06–0.07 s^{-1}. In all of these experiments, irrespective of the application, only minor amounts of Pd, typically less than 2.5%, were leached into the reaction mixture [49].

Grafting of another chelating ligand was accomplished via the ROMP of 4′-(norborn-2-en-5-ylmethylenoxy)terpyridine. The polymerization of this monomer by Mo(N-2,6-i-Pr$_2$ – C$_6$H$_3$)(CHCMe$_2$h)(CMe(CF$_3$)$_2$)$_2$ only fulfills the requirements of a class V living system [7]. Consequently, the corresponding surface-grafted support had to be prepared using a "grafting-from" approach, as described above [38] (Scheme 9).

Loading with Cu (I) afforded the desired ATRP support [50–52]. Typical metal loadings were 15 mmol/g. Polystyrene (PS) prepared under ATRP conditions with these supports showed comparatively low polydispersities (PDI = 1.55–1.77). The ATRP system consisted of a metal center with one terpyridyl and presumably three acetonitrile ligands, which were (at least partly) substituted by the monomer. Consequently, and in contrast to standard systems [53], the equilibrium M^{n+} ↔ M^{n+1} in this type of reaction did not require conformational changes or dissociation of a terpyridyl ligand. Therefore, polymerization proceeded comparably fast: within two hours. Unfortunately, presumably to unfavorable equilibria involving the dormant species, polymer yields were low (< 35%).

The ability of ROMP to polymerize even more complex functional monomers was demonstrated by the fact that the cationic NHC precursor 1,3-di(1-mesityl)-4-{[(bicyclo[2.2.1]hept-5-en-2-ylcarbonyl)oxy]methyl}-4,5-dihydro-1H-imidazol-3-ium tetrafluoroborate can be polymerized using both ruthenium- and molybdenum-based initiators. Thus, reaction of this monomer with RuCl$_2$(PCy$_3$)$_2$(CHPh) in methylene chloride at 45 °C results in com-

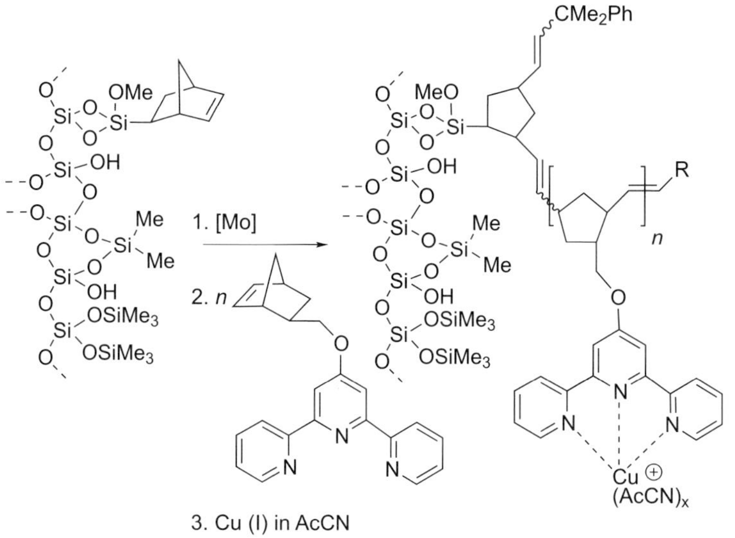

Metathesis Polymerization To and From Surfaces 151

Scheme 9 Grafting of 4′-(norborn-2-en-5-ylmethylenoxy)terpyridine on silica and loading with Cu(I)

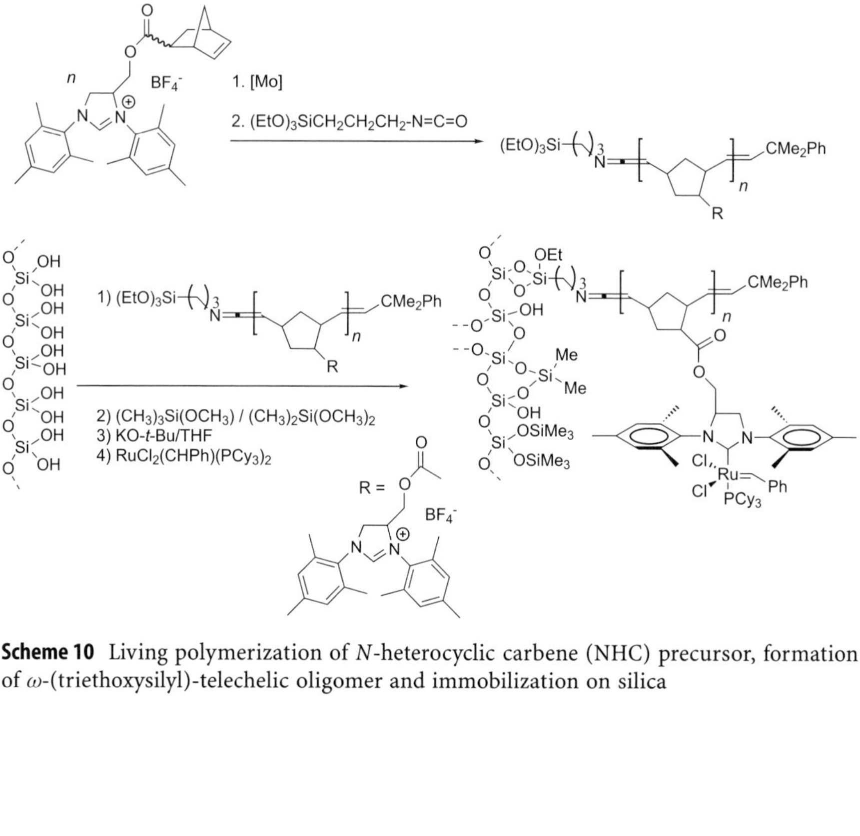

Scheme 10 Living polymerization of N-heterocyclic carbene (NHC) precursor, formation of ω-(triethoxysilyl)-telechelic oligomer and immobilization on silica

plete consumption of the initiator and formation of an oligomer with a DP of 7. Alternatively, polymerization can be carried out with the Schrock initiator Mo(*N*-2,6-*i*-Pr$_2$ – C$_6$H$_3$)(CHCMe$_2$Ph)(OCMe(CF$_3$)$_2$)$_2$ [54] at ambient temperature in methylene chloride. The observed theoretical DP of 7 was in excellent agreement with a DP of 7 ± 1 found via endgroup analysis using ^{1}H NMR. The polymerization system fulfills the requirements of a class V living polymerization system at least, which allows quantitative conversion into a telechelic polymer. Thus, an endgroup suitable for grafting onto silica was introduced by reacting the living polymer in a Wittig-type reaction with an excess of ω-(triethoxysilyl)propylisocyanate (Scheme 10).

Next, telechelic oligo-(1,3-di(1-mesityl)-4-{[(bicyclo[2.2.1]hept-5-en-2-ylcarbonyl)oxy]methyl}-4,5-dihydro-1*H*-imidazol-3-ium tetrafluoroborate) was reacted with silica-60. Reaction of the grafted supports with KO-*t*-Bu in THF at – 30 °C yielded the free carbene, which was subsequently reacted with RuCl$_2$(PCy$_3$)$_2$(CHPh) to yield the immobilized second-generation Grubbs catalyst [55]. The ruthenium content of the solution, as measured by inductively coupled plasma-optical emission spectroscopy (ICP-OES), revealed catalyst loadings of 0.1–0.5 wt. %. Ring-closing metathesis (RCM) carried out

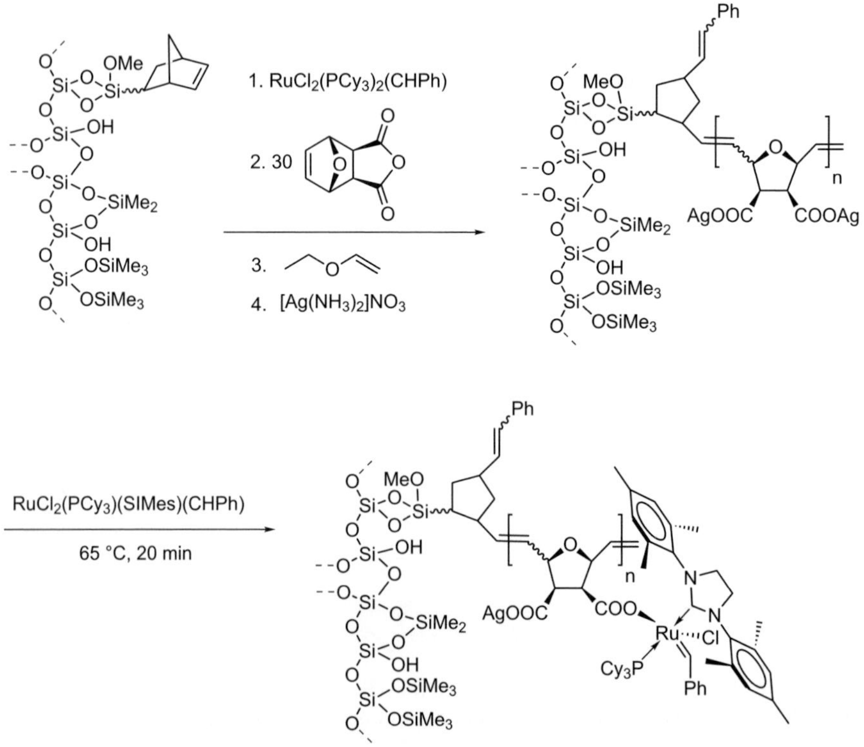

Scheme 11 Synthesis of a silica-immobilized Grubbs-Herrmann catalyst

with diethyl diallylmalonate (DEDAM) as a benchmark gave TONs ≤ 80 for a stirred batch. No catalyst bleeding was observed, thus offering access to virtually metal-free products.

Norborn-2-ene-5-yl-trichlorosilane or norborn-2-ene-5-yl-triethoxysilane) (both *exo/endo*-mixtures) surface-derivatized silica was consecutively reacted with $RuCl_2(PCy_3)_2(CHPh)$ and *exo,exo-7*-oxanorborn-2-ene-5,6-dicarboxylic anhydride and 7-oxanorborn-2-ene-5-carboxylic acid. With this "grafting-from" approach, satisfactory amounts of both monomers were grafted onto the support. Thus, anhydride loadings of 0.22 mmol/g (LiChrospher 300-5) and 1.2 mmol/g (Nucleosil 300-7) were achieved. Conversion into the corresponding di- and mono-silver salts and reaction with $RuCl_2(PCy_3)(IMesH_2)(CHPh)$ gave the desired supported catalysts (Scheme 11).

The catalyst-loading was in the range of 42 mg (Nucleosil 300-7) to 63 mg catalyst/g (Lichrospher). RCM reactions carried out with these two silica-supported catalyst versions allowed TONs of up to 90 for a series of simple α,ω-dienes [56].

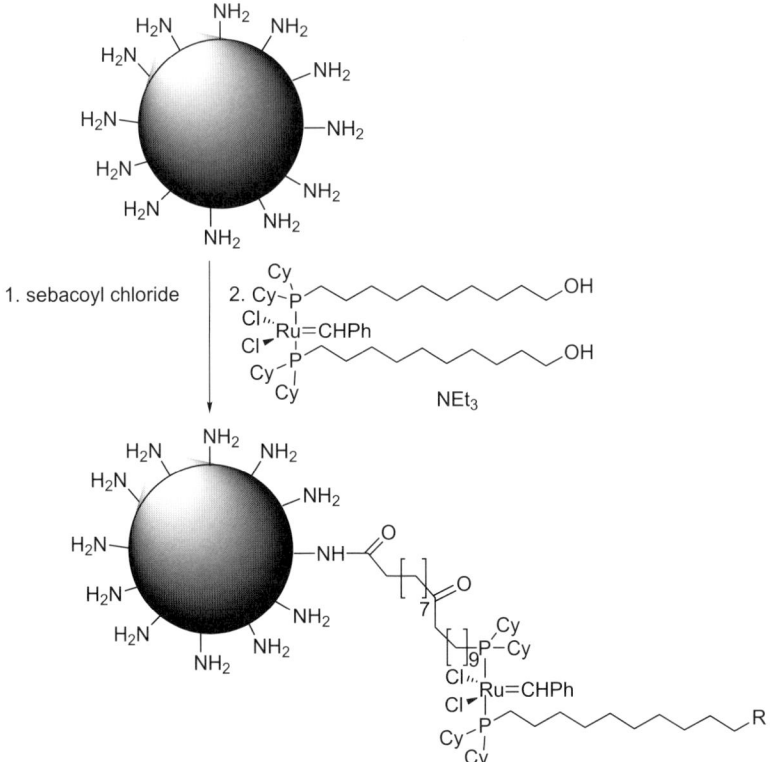

Scheme 12 Preparation of hybrid core-shell particles

Identical protocols for the preparation of surface-bound thin polymer films using Si/SiO$_2$ surface-bound norborn-5-ene-2-ylsilanes were described by other groups [57]. Combining a "grafting-from" approach with microcontact imprinting, patterned polymer films of variable thickness (5–500 nm) consisting of poly(5-triethoxysilylnorborn-5-ene) were prepared using RuCl$_2$(PCy$_3$)$_2$(CHPh) as initiator. Lateral dimensions as small as 2 μm could be realized [58]. Recently, Mingotaud et al. reported on the immobilization of the first-generation Grubbs' type catalyst RuCl$_2$(PCy$_3$R)$_2$(CHPh) (R = (CH$_2$)$_{10}$ – OH) on amino-functionalized silica using sebacoylchloride. The immobilized system was used for the preparation of hybrid core-shell particles using norborn-2-ene as monomer [59] (Scheme 12).

3.1.3
Other Inorganic Surfaces

Emrick and Coughlin et al. reported on the synthesis of cadmium selenide–polymer composites. A vinylbenzyl-derivatized phosphine oxide was physisorbed onto cadmium selenide particles. Subsequent reaction with RuCl$_2$(PCy$_3$)$_2$(CHPh) or RuCl$_2$(NHC)(PCy$_3$)(CHPh) (NHC = 1,3-dimesitylimidazol-2-ylidene) followed by addition of cyclooctene, 7-oxanorborn-5-ene-2,3-dicarboxylic anhydride, dicyclopentadiene or N-methyl-7-oxanorborn-5-

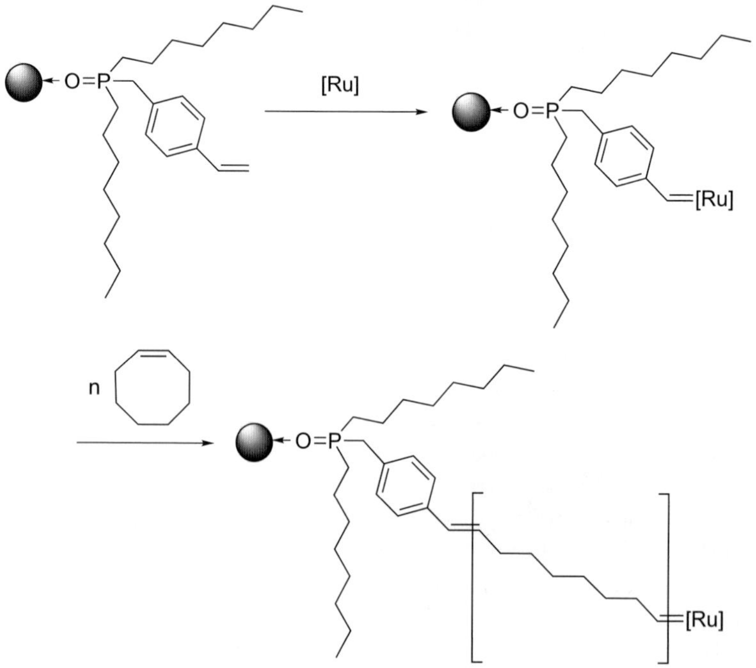

Scheme 13 Synthesis of cadmium selenide–polymer composites

ene-2,3-dicarboxyimide resulted in the desired surface modification and formation of the composite, which were suggested to possess interesting solution and electronic properties (Scheme 13) [60].

3.2
Organic Surfaces

3.2.1
Merrifield-Type Resins

Merrifield-type resins are still among the most prominent organic supports in both organic synthesis and catalysis. The most straightforward methods for providing anchoring groups for the subsequent attachment of other groups or polymers are the chloromethylation of PS-DVB or the substitution of styrene by chloromethylstyrene during synthesis of these supports. However, bromomethyl groups appear favorable, since they exhibit enhanced reactivity compared to their chloromethyl analogs. They can be generated by bromomethylation using trioxane, tin tetrabromide and trimethylbromosilane [61] or by conversion of the chloromethyl groups into the corresponding bromomethyl groups via halogen exchange [62, 63]. To graft using ROMP, the bromomethylated poly(styrene-co-divinylbenzene) PS-DVB resins can be converted into the norborn-2-ene-5-ylmethylethers via standard Williamson ether synthesis. Up to 2 mmol/g of norborn-5-ene-5,6-dicarboxylic anhydride can be grafted onto standard norborn-5-ene-2-yl-derivatized Merrifield resins (2% crosslinked) using either a "grafting-from" or a "grafting-to" approach.

Alternative supports to Merrifield resins (the synthesis of ROMP spheres for use in combinatorial chemistry) was reported by Barrett and coworkers [64]. Here, surface functionalization was accomplished via reaction of vinyl-PS-DVB, with a low degree of crosslinking, with $RuCl_2(PCy_3)_2$ (CHPh) to form the immobilized catalytic species. Reaction with a functional monomer, such as norborn-2-en-5-ylmethyl-4-bromobenzoate, gave the corresponding support with loadings of up to 3 mmol of functional monomer/g resin (Scheme 14) [65]. Swelling properties were similar to materials prepared by ring-opening metathesis precipitation polymerization reported by our group [48, 66–69].

Recently, Caster et al. described the surface modification of multifilament fibers such as nylon or Kevlar [70]. Coating techniques using preformed ROMP-based polymers and process contact metathesis polymerization (CMP), initially described by Grubbs et al. [71], were both used. The latter involves a procedure where the initiator is physisorbed onto the surface of a substrate and fed with a ROMP-active monomer that finally encapsulates the substrate. These modified fibers showed improved adhesion to natural rubber elastomers.

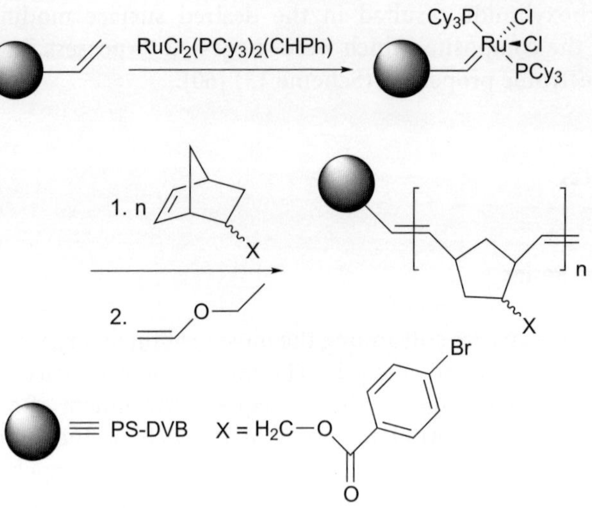

Scheme 14 Surface grafting of organic supports starting from surface-immobilized vinyl groups

3.2.2
Monolithic Supports

Monolithic separation media evolved as a successful joint venture between the materials and separation sciences. Based on theoretical reflections, the common idea was to produce a support with a high degree of continuity that would meet the requirements for fast yet highly efficient separations [72, 73]. The first experiments in this direction were carried out in the 1960s and 1970s [74, 75], but it took around twenty years to adapt this new technology to the area of heterogeneous catalysis. Meanwhile, these supports, usually referred to as monolithic supports, continuous beds or rigid rods [75] were successfully applied to liquid chromatography (including microseparation) [76–79], capillary electrochromatography, as well as in solid phase extraction (SPE) [80]. In these separation techniques, the focus was on both medium and high molecular mass biopolymers [81], and even on low molecular mass analytes [82–85]. Quite recently (in 2001), our group developed a ROMP-based synthesis for these types of materials and described the use of these supports in separation science and heterogeneous catalysis [49, 86–98].

3.2.2.1
Basics and Concepts

Generally speaking, the term "monolith" applies to any single-body structure containing interconnected repeating cells or channels. Such materials may either be metallic or prepared from inorganic mixtures (for example by a sintering process to form ceramics [99]), or from organic compounds (usually via crosslinking polymerization [100, 101]). Here, the term "monolith" or "rigid rod" will be used for crosslinked, organic materials which are characterized by a defined porosity and which support interactions/reactions between this solid and the surrounding liquid phase. Besides advantages such as decreased back-pressure and enhanced mass transfer [102, 103], the ease of fabrication as well as the many structural alteration possibilities available should be mentioned.

A considerable variety of functionalized and nonfunctionalized monolithic materials based on either organic or inorganic polymers are currently available. While inorganic monoliths are usually prepared from silica precursors ($Si(OR)_4$) via sol-gel techniques [82, 84, 85], organic continuous beds are usually prepared from methacrylates or poly(styrene-co-divinylbenzene) [100, 104–107], almost always by free radical polymerization. Profound insights into the field of sol-gel and free radical polymerization-based monoliths can be found in books dedicated to this subject [108]. Despite the comparably poor control over free radical polymerization-based systems, the porosity and microstructure of monolithic materials has been successfully varied [100]. In contrast to the free radical process, our group confirmed the general applicability of a transition metal-based polymerization technique such as ROMP to the synthesis of high-performance monolithic separation media. Due to this broad applicability of ROMP and the good definition of the resulting materials, we investigated the extent to which this transition metal-catalyzed polymerization could be used to synthesize monolithic polymers [87]. We found that this may be accomplished by generating a continuous matrix by ring-opening metathesis copolymerization of suitable monomers with a crosslinker in the presence of porogenic solvents within a device (column). In addition, we elaborated concepts for the surface grafting of these monolithic supports.

3.2.2.2
Manufacture of ROMP-Based Monolithic Supports

The choice of a suitable initiator represents an important step in creating a well-defined polymerization system in terms of initiation efficiency and control over propagation. The entire system can only be designed on a *stoichiometric base* when a quantitative and fast initiation occurs. This is of enormous importance, because the composition of the entire polymerization mixture needs to be varied within small increments in order to control the microstructure. The catalyst needs to be carefully selected from both chemical and practical points of view. Schrock [5, 10, 12, 109, 110] and Grubbs systems [6], both highly active in the ROMP of strained functionalized olefins, can often be used. Since the preparation and in particular derivatization of ROMP-based rigid rods requires some handling that is difficult to perform in an inert atmosphere, the less oxygen-sensitive and less reactive ruthenium-based Grubbs-type initiators were used. Since $RuCl_2(PPh_3)_2(CHPh)$ turned out to be too unreactive, $RuCl_2(PCy_3)_2(CHPh)$ was used. In order to avoid any uncontrolled, highly exothermic reactions, neither the Grubbs-Herrmann nor the Grubbs-Hoveyda catalyst were applied. Among the various possible combinations of monomers and crosslinkers, such as norborn-2-ene, norbornadiene (NBDE), dicyclopentadiene (DCPD), 1,4,4a,5,8,8a-hexahydro-1,4,5,8-*exo-endo*-dimethanonaphthalene (DMN-H6), tris(norborn-2-en-5-ylmethylenoxy)methylsilane (NBE-CH$_2$O)$_3$SiCH$_3$ and 1,4a,5,8,8a,9,9a,10,10a-decahydro-1,4,5,8,9,10-trimethanoanthracene, the copolymerization of norborn-2-ene with DMN-H6 or (NBE-CH$_2$O)$_3$SiCH$_3$ in the presence of two porogenic solvents (2-propanol and toluene) with $RuCl_2(PCy_3)_2(CHPh)$ worked best (Scheme 15).

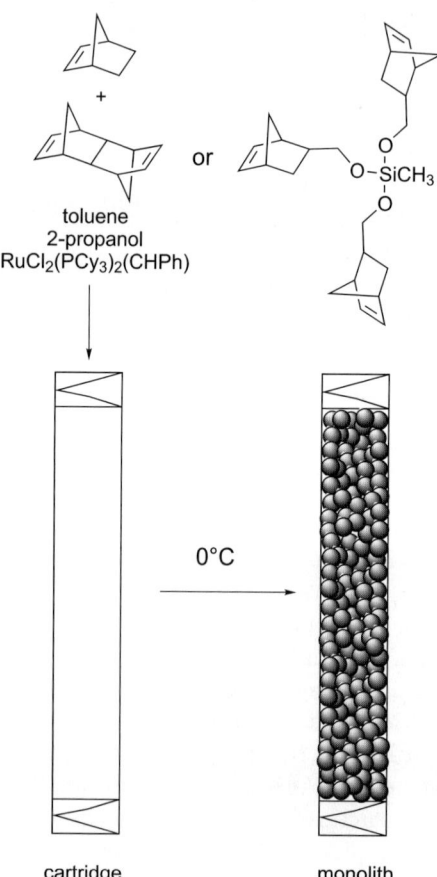

Scheme 15 Synthesis of a ROMP-derived monolith

3.2.2.3
Microstructure of Metathesis-Based Rigid Rods

In order to understand monolithic supports and the effects of polymerization parameters, a brief description of the general construction of a monolith in terms of microstructure, backbone and relevant abbreviations is given in Fig. 5 [86, 87]. As can be deduced from this, monoliths consist of interconnected microstructure-forming microglobules, which are characterized by a certain diameter (d_p) and microporosity (ε_p). In addition, the monolith is characterized by an inter-microglobule void volume (ε_z), which is mainly responsible for the back-pressure at a certain flow rate.

The volume fractions of both micropores (ε_p) and voids (intermicroglobule porosity, ε_z) represent the total porosity (ε_t). This value indicates the percentage of pores in the monolith. The pore size distribution can be calculated from inverse size exclusion chromatography (ISEC) data [111, 112] or from mercury intrusion [113]. Together, these two values directly translate into the total pore volume V_p, expressed in mL/g. In addition, one may calculate the specific surface area σ, expressed in m^2/g, from these. In order to design monolithic supports for different tasks, the influence of all of the variables – the components of the polymerization mixture (NBE, DMN-H6 or NBE-CH$_2$O)$_3$SiCH$_3$, the solvents, the free phosphine and the initiator as well as temperature – on microstructure formation was investigated. The relative ratios of all of the components – NBE, DMN-H6, porogens and catalyst – allowed broad variations in the microstructure of the monolithic material, including structures ideal for heterogeneous catalysis. In summary, the volume

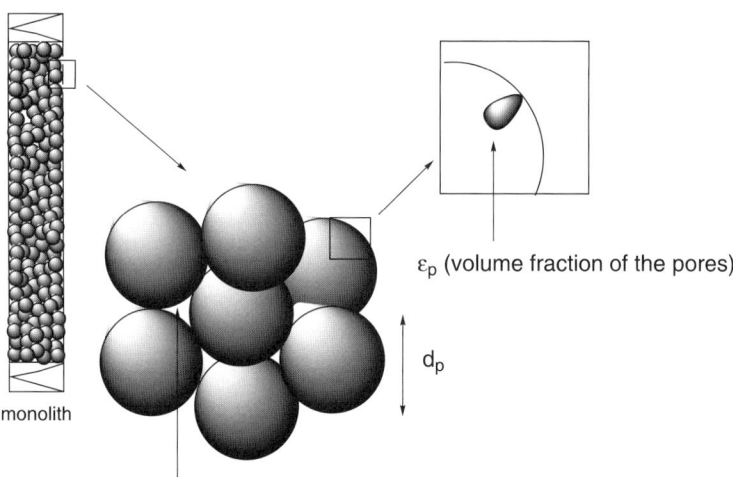

Fig. 5 Construction of a monolith

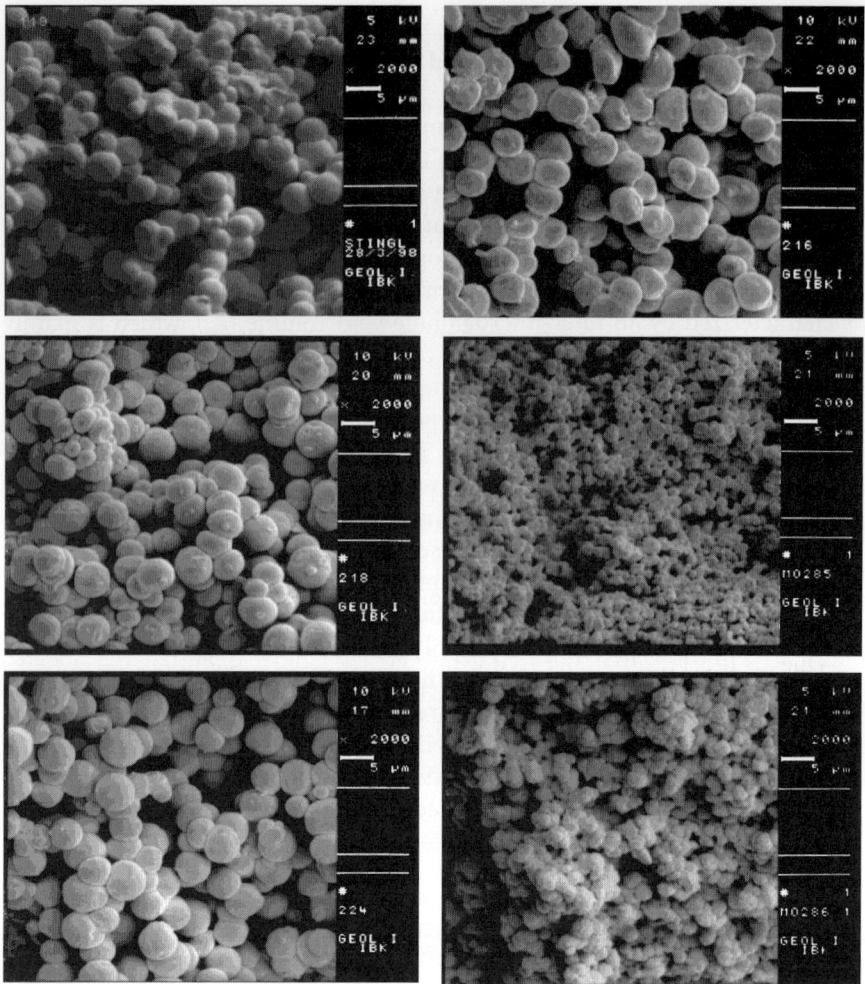

Fig. 6 Structural variations in ROMP-derived monoliths

fraction of the interglobular void volume (ε_z) and the total porosity (ε_t) were varied within 0–50% and 50–80%, respectively. Figure 6 illustrates some of the microstructures that were generated.

3.2.2.4
Functionalization and Metal Removal

Functionalization can be conveniently achieved using the ROMP-based protocol. Thus, the "living" character [7, 114–118] of ruthenium-catalyzed polymerization offers a perfect route to functionalization. In fact, the active ruthenium sites can be used for derivatization after rod formation is complete. ICP-OES-based investigations revealed that more than 98% of the *initial amount* of initiator is located at the microglobule *surface* once microstructure formation is complete [89]. These findings can be explained by the highly polar character of the initiator, which preferably locates at the interface between the apolar, toluene-enriched microglobule and polar, 2-propanol-enriched interface. Using the initiator covalently bound to the surface, a series of functional monomers can be grafted onto the monolith surface by simply passing solutions through the mold (Scheme 16) [86, 87].

Since no crosslinking can take place, tentacle-like polymer chains attached to the surface are formed. In addition, microglobules are designed in such a way that their pore size is < 1.2 nm, which restricts functionalization to their surface [93, 94]. The degree of this graft polymerization of functional monomers varies over almost two orders of magnitude, depending on their ROMP activity. This "in situ, grafting-from" approach offers multiple advantages. First, the structure of the parent monolith is not affected by the functional monomer and can be optimized regardless of the functional monomer used later. Second, solvents other than the porogens (such as methylene chloride or DMF) may be used for the "*in situ*" derivatization, depending on the solubility of the monomer. An overview of the different monomers that have already been grafted is given in Table 2.

Due to the fact that the living initiator is almost quantitatively located at the surface of the microglobules, the efficiency of *metal removal* from the monolith after polymerization is high. Investigations revealed that the remaining ruthenium concentration after capping with ethyl vinyl ether is below 10 µg/g, corresponding to a metal removal of more than 99.8%.

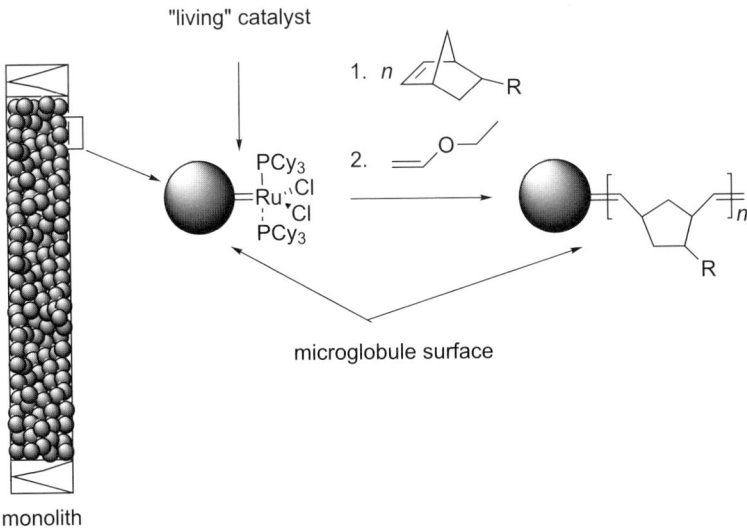

Scheme 16 In situ functionalization of a ROMP-derived monolith

Table 2 Functional monomers used for in situ grafting of ROMP-derived monoliths

monomer	mmol/g
(norbornene dicarboxylic anhydride, epoxide)	0.2
(epoxy-norbornene ~COOH)	0.14
(norbornene dicarboximide with valine amide-NO₂ aryl)	0.03
(norbornene carboxylate ester with cyclodextrin); X = O, CH₂	–
(norbornene dicarboximide-N-C₆H₄-NMe₂)	0.26
(norbornene dicarboximide-N-phenyl)	0.22
(norbornene dicarboximide-N-C₆H₄-OH)	0.06
(norbornene-COO-imidazolinium BF₄⁻; R–N⁺=N–R)	0.002

Deleuze et al. used the same approach for the synthesis and functionalization of emulsion-derived polymeric foams [119]. Alternatively, a post-synthesis grafting method recently developed in our group offers access to high-capacity functionalized monolithic systems. Such high capacity monoliths are vital to various applications such as catalysis, extraction of environmental contaminants, extraction of biochemicals for either pharmaceutical or clinical purposes or, more generally, separation techniques [100]. With these systems, amounts of grafted monomers can exceed 1 mmol/g [94].

3.2.2.5
Applications of Functionalized Metathesis-Based Monoliths to Catalysis

In heterogeneous catalysis, one wants to combine the general advantages of homogeneous systems, such as high definition and activity, with the advantages of heterogeneous catalysis, such as increased stability, ease of separation, and recycling. So far, monolithic catalytic media have largely been restricted to metal oxides, porous metals and certain polysaccharides [120]. The first successful use of metathesis-based monolithic media for heterogeneous catalysis was accomplished by using these supports as carriers for Grubbs-type initiators based on *N*-heterocyclic carbenes (NHCs) [121, 122]. In order to generate sufficient porosity, monoliths with a suitable microporosity (40%) and microglobule diameter (1.5 ± 0.5 μm) were synthesized. Consecutive in situ derivatization was successfully accomplished using a mixture of norborn-2-ene and a polymerizable NHC-precursor (Scheme 17) [55, 89, 123–125].

The use of norborn-2-ene drastically enhanced grafting yields for the functional monomer. Using this set-up, tentacles of copolymer with a degree of oligomerization of the functional monomer of 2–5 were generated. The free NHC necessary for catalyst formation was simply generated using a strong

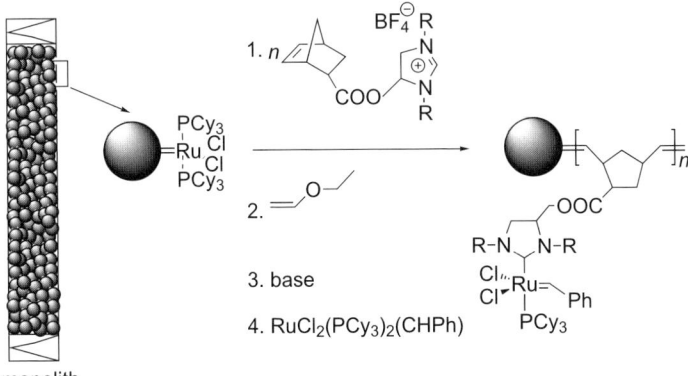

Scheme 17 Immobilization of a NHC precursor on a ROMP-derived monolith via in situ grafting

base such as 4-dimethylaminopyridine (DMAP). In a last step, excess base was removed by extensive washing and the catalyst was immobilized/formed by passing a solution of $RuCl_2(PCy_3)_2(CHPh)$ over the rigid rod. Loadings of up to 1.4% of Grubbs-catalyst on NHC base were achieved. Monolith-immobilized metathesis catalysts prepared by this approach showed high activity in various metathesis-based reactions such as ROMP and RCM. The *cis/trans* ratios of polymers (90%) exactly corresponded to the ones found with analogous homogeneous systems. The use of chain-transfer agents (CTAs, such as *cis*-1,4-diacetoxybut-2-ene, diethyl diallylmalonate, 2-hexene) allowed the regulation of molecular mass, particularly in the case of cyclooctene. The presence of CTAs also enhanced the lifetime of the catalytic centers by reducing the average lifetime of the ruthenium methylidenes, thus allowing the prolonged use of these systems. Additionally, both the tentacle-type structure and the designed microstructure of the support reduced diffusion to a minimum. In a benchmark reaction with DEDAM, these properties directly translated into high average turnover frequencies (TOFs) of up to 0.5 s^{-1}.

Alternatively, monolith-supported second-generation Grubbs catalysts containing saturated (such as IMes = 1,3-dimesitylimidazol-2-ylidene) or saturated (such as SIMes = 1,3-dimesitylimidazolin-2-ylidene) NHCs [4] were prepared by a synthetic protocol summarized in Scheme 16. Surface-derivatization of a monolith was achieved with 7-oxanorborn-5-ene-2,3-dicarboxylic anhydride followed by conversion of the grafted poly(anhydride) into the corresponding poly-silver salt. This silver salt was used for halogen exchange with a broad variety of second-generation Grubbs catalysts, leading to the catalytic species shown in Scheme 18. In a benchmark reaction with DEDAM, TONs of up to 830 were achieved [56, 126, 127].

Finally, a monolith-supported version of the Grubbs-Hoveyda catalyst was prepared in an analogous approach using a perfluoroglutaric anhydride-derived ligand (Scheme 19). When used in continuous flow experiments, TOFs of 0.1 s^{-1} were observed, and TONs were > 500 [127, 128].

All of the monolith-based catalytic systems summarized here were successfully used as pressure-stable catalytic reactors. Bleeding was suppressed, leading to virtually ruthenium-free products (ruthenium contents of far below 0.1%), even in RCM [129].

A "grafting-from" approach was also used to immobilize dipyridylamide ligands. Nevertheless, due to the incompatibility of the ruthenium initiator with pyridine-containing ligands employed, a completely new approach was elaborated for monolith functionalization with these ligands. Since the nitrogen lone pair has to be protected in order to prevent coordination to the ruthenium core, the complex *N,N*-dipyrid-2-yl-7-oxanorborn-5-en-2-ylcarbamido palladium dichloride was synthesized and grafted onto a monolith using norborn-2-ene as a comonomer (Scheme 20).

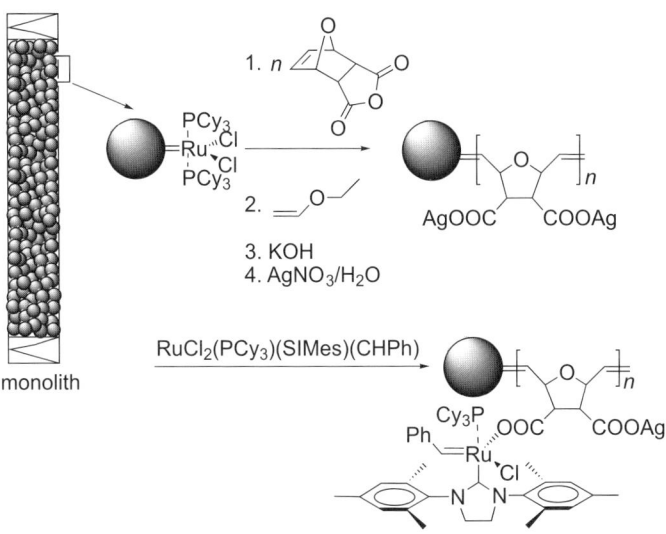

Scheme 18 Immobilization of a metathesis catalyst on a ROMP-derived monolithic support

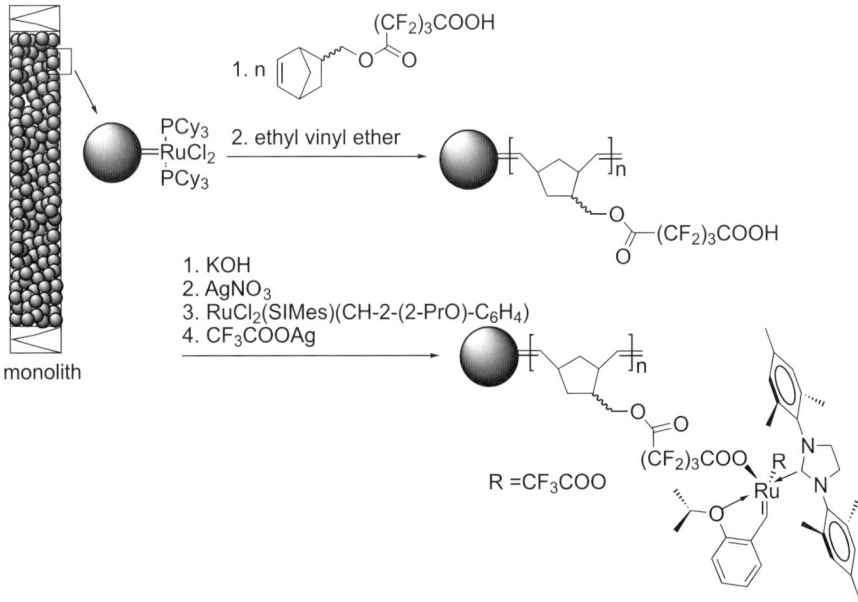

Scheme 19 Immobilization of a metathesis catalyst on a ROMP-derived monolithic support using fluorinated carboxylates

Following this "grafting-from" approach, monoliths containing 7 μmol (0.07%) Pd were prepared. In a model reaction between styrene and iodobenzene, TOFs of 1.2–1.6 s^{-1} were found. These figures clearly exceed those

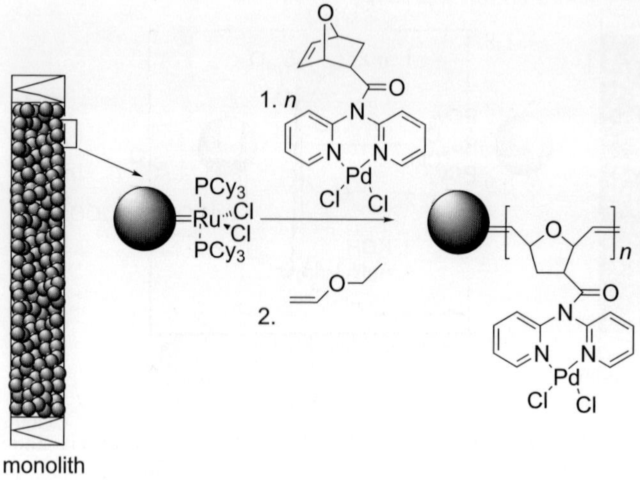

Scheme 20 Immobilization of a Heck catalyst on a ROMP-derived monolithic support

obtained with supports prepared by ring-opening metathesis precipitation polymerization, which gave TOFs of 0.35 s^{-1} in identical reactions [130]. Pd leaching was generally low (2.2% in total for over ten hours). Alternatively, if used as a cartridge, typical amounts of reactants used in combinatorial chemistry (50–100 mg in total) were converted in satisfactory yields (< 80%) [49].

3.2.2.6
Applications of Functionalized Metathesis-Based Monoliths to Separation Science

Due to the pure hydrocarbon backbone, monoliths prepared from NBE and DMN-H6 are strongly hydrophobic. Impressive separation abilities have been demonstrated by the fast separation of biologically relevant compounds such as proteins, double-stranded (ds) DNA, oligonucleotides and phosphorothioate oligodeoxynucleotides [45, 86–88, 90, 92]. While non-functionalized monolithic media are useful supports for the separation of biomolecules via a reversed-phase (RP-) mechanism, functionalized analogs are highly attractive for other separation problems that could be solved by other separation mechanisms, such as by chiral chromatography or ion chromatography. Using a β-cyclodextrin-grafted monolith, proglumide (a β-blocker) was separated within three minutes, which clearly underlines the potential of this type of chemistry in many areas of separation science [87]. In addition, a mixture of oligomeric desoxythymidylic acids – dT_8, dT_{12} and dT_{24} – were separated on a poly(N-(N,N-dimethylaminoprop-1-yl)norborn-5-ene-2,3-dicarboxylic imide)-grafted ROMP-derived monolith (Fig. 7).

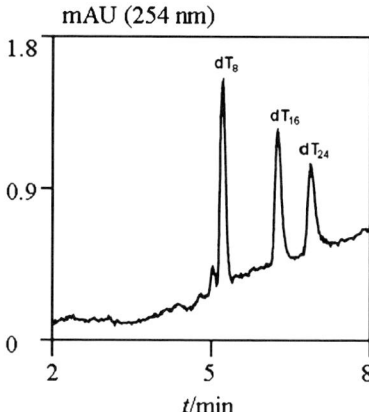

Fig. 7 Separation of dT_8, dT_{12}, and dT_{24} on a ROMP-based monolith via anion-exchange chromatography. Monolith (60 × 0.2 mm), gradient: ammonium acetate (20 mM)-20% acetonitrile, pH 5 to Tris-HCl (20 mM)-20% acetonitrile, pH 8, salt gradient from 0–1 M NaCl in 10 min, 4.3 µl/min

3.2.3
Other Organic Surfaces

Dai et al. reported on the functionalization of carbon nanotubes. Vinyl- and norborn-2-ene-functionalized pyrenes, respectively, were adsorbed onto the surfaces of the nanotubes and used to graft norborn-2-ene [131]. $RuCl_2(PCy_3)_2(CHPh)$ was used throughout. A polymer film thickness of between 5 and 20 nm was achieved through this approach.

4
Conclusion/Outlook

Metathesis-based polymerization techniques have certainly found their place in materials science. This has been made possible by adding well-defined and tolerant initiators to the armor of existing polymerization systems. Using these initiators, ROMP and 1-alkyne polymerization have had an enormous impact on the development of surface-modified organic and inorganic materials. Applications to catalysis and separation science have occurred as well as in the more "traditional" fields of optics and electronics. The ongoing developments in organometallic chemistry, polymer chemistry, and in particular in metathesis polymerization will undoubtedly result in permanent improvements of existing systems and techniques as well as in new applications in many areas of chemistry and materials science.

Acknowledgements Our work was supported in part by grants provided by the *Austrian Science Fund* (projects P-11740-GEN, P-12963-GEN, START Y-158), the *Österreichische Nationalbank* (project 7489), the *University of Innsbruck*, the *Industriellenvereinigung Tirol* and the *European Commission* (contracts F14W-CT96-0019 and FIS 1999-00130). I wish to thank all undergraduate, graduate, and postgraduate students involved in the work described in this chapter for their dedication and enthusiasm.

References

1. Buchmeiser M (1997) Macromolecules 30:2274
2. Buchmeiser MR, Schuler N, Kaltenhauser G, Ongania K-H, Lagoja I, Wurst K, Schottenberger H (1998) Macromolecules 31:3175
3. Buchmeiser MR, Schuler N, Schottenberger H, Kohl I, Hallbrucker A (2000) Des Monomers Polymers 3:421
4. Buchmeiser MR (2000) Chem Rev 100:1565
5. Schrock RR (1994) Pure Appl Chem 66:1447
6. Trnka TM, Grubbs RH (2001) Acc Chem Res 34:18
7. Matyjaszewski K (1993) Macromolecules 26:1787
8. Krause JO, Zarka MT, Anders U, Weberskirch R, Nuyken O, Buchmeiser MR (2003) Angew Chem 115:6147
9. Schrock RR (1993) In: Brunelle DJ (ed) Ring-opening polymerization. Hanser, Munich, p 129
10. Schrock RR (1995) Polyhedron 14:3177
11. Ivin KJ, Mol JC (1997) Olefin metathesis and metathesis polymerization. Academic, San Diego, CA
12. Schrock RR (2002) Chem Rev 102:14
13. Frenzel U, Nuyken O (2002) J Polym Sci Polym Chem 40:2895
14. Frenzel U, Weskamp T, Kohl FJ, Schattenmann WC, Nuyken O, Herrmann WA (1999) J Organomet Chem 586:263
15. Weskamp T, Schattenmann WC, Spiegler M, Herrmann WA (1998) Angew Chem 110:2631
16. Weskamp T, Kohl FJ, Herrmann WA (1999) J Organomet Chem 582:362
17. Weskamp T, Kohl FJ, Hieringer W, Gleich D, Herrmann WA (1999) Angew Chem 111:2573
18. Garber SB, Kingsbury JS, Gray BL, Hoveyda AH (2000) J Am Chem Soc 122:8168
19. Van Veldhuizen JJ, Garber SB, Kingsbury JS, Hoveyda AH (2002) J Am Chem Soc 124:4954
20. Kingsbury JS, Harrity JPA, Bonitatebus PJ Jr, Hoveyda AH (1999) J Am Chem Soc 121:791
21. Bielawski CW, Grubbs RH (2000) Angew Chem 112:3025
22. Lynn DM, Kanaoka RH, Grubbs RH (1996) J Am Chem Soc 118:784
23. France MB, Grubbs RH, McGrath DV, Paciello RA (1993) Macromolecules 26:4742
24. Mohr B, Lynn DM, Grubbs RH (1996) Organometallics 15:4317
25. Lynn DM, Mohr B, Grubbs RH (1998) J Am Chem Soc 120:1627
26. Lynn DM, Mohr B, Grubbs RH (1998) Polym Prepr 39:278
27. Lynn DM, Mohr B, Grubbs RH, Henling LM, Day MW (2000) J Am Chem Soc 122:6601
28. Watson KJ, Zhu J, Nguyen ST, Mirkin CA (1999) J Am Chem Soc 121:462

29. Watson KJ, Zhu J, Nguyen ST, Mirkin CA (2000) Pure Appl Chem 72:67
30. Liu X, Guo S, Mirkin CA (2003) Angew Chem 115:4933
31. Weck M, Jackiw JJ, Rossi RR, Weiss PS, Grubbs RH (1999) J Am Chem Soc 121:4088
32. Rutenberg IM, Scherman OA, Grubbs RH, Jiang W, Garfunkel E, Bao Z (2004) J Am Chem Soc 126:4062
33. Li X-M, Huskens J, Reinhoudt DN (2003) Nanotechnology 14:1064
34. Wu M, O'Neill SA, Brousseau LC, McConnell WP, Shultz DA, Linderman RJ, Feldheim DL (2000) Chem Commun 775
35. Samanta D, Faure N, Rondelez F, Sarkar A (2003) Chem Commun 1186
36. Juang A, Scherman OA, Grubbs RH, Lewis NS (2001) Langmuir 17:1321
37. Harada Y, Girolami GS, Nuzzo RG (2003) Langmuir 19:5104
38. Buchmeiser MR, Sinner F, Mupa M, Wurst K (2000) Macromolecules 33:32
39. Buchmeiser MR, Sinner FM (1999) Austrian Patent Application AT 070499
40. Buchmeiser MR (2002) In: Khosravi E, Szymanska-Buzar T (eds) NATO Science Series II. Mathematics, physics and chemistry, vol 56. Klywer, Dordrecht, p 205
41. Blümel J (1995) J Am Chem Soc 117:2112
42. Behringer KD, Blümel J (1996) J Liq Chromatogr R T 19:2753
43. Mayr B, Sinner F, Buchmeiser MR (2001) J Chromatogr A 907:47
44. Mayr B, Schottenberger H, Elsner O, Buchmeiser MR (2002) J Chromatogr A 973:115
45. Mayr B, Buchmeiser MR (2001) J Chromatogr A 907:73
46. Buchmeiser MR, Schrock RR (1995) Macromolecules 28:6642
47. Eder K, Reichel E, Schottenberger H, Huber CG, Buchmeiser MR (2001) Macromolecules 34:4334
48. Sinner F, Buchmeiser MR, Tessadri R, Mupa M, Wurst K, Bonn GK (1998) J Am Chem Soc 120:2790
49. Buchmeiser MR, Lubbad S, Mayr M, Wurst K (2003) Inorg Chim Acta 345:145
50. Buchmeiser MR, Kröll R, Wurst K, Schareina T, Kempe R, Eschbaumer C, Schubert US (2001) In: Adler H-JP, Arndt K-F, Kuckling D, Stamm M, Voit B, Wolff T (eds) Makromol Symp (Tailormade polymers) 164:187
51. Kröll R, Eschbaumer C, Schubert US, Buchmeiser MR, Wurst K (2001) Macromol Chem Phys 202:645
52. Schubert US, Weidl CH, Eschbaumer C, Kröll R, Buchmeiser MR (2001) Polym Mater Sci Eng 84:514
53. Matyjaszewski K, Xia J (2001) Chem Rev 101:2921
54. Oskam JH, Fox HH, Yap KB, McConville DH, O'Dell R, Lichtenstein BJ, Schrock RR (1993) J Organomet Chem 459:185
55. Mayr M, Buchmeiser MR, Wurst K (2002) Adv Synth Catal 344:712
56. Krause JO, Lubbad S, Nuyken O, Buchmeiser MR (2003) Adv Synth Catal 345:996
57. Kim NY, Jeon NL, Choi IS, Takami S, Haada Y, Finnie KR, Girolami GS, Nuzzo RG, Whitesides GM, Laibinis PE (2000) Macromolecules 33:2793
58. Jeon Nl, Choi IS, Whitesides GM, Kim NY, Laibinis PE, Harada Y, Finnie KR, Girolami GS, Nuzzo RG (1999) Appl Phys Lett 75:4201
59. Mingotaud A-F, Reculusa S, Mingotaud C, Keller P, Sykes C, Duguet E, Ravaine S (2003) J Mater Chem 13:1920
60. Skaff H, Ilker MF, Coughlin EB, Emrick T (2002) J Am Chem Soc 124:529
61. Itsuno S, Uchikoshi K, Ito K (1990) J Am Chem Soc 112:8187
62. Yoon KB, Kochi JK (1987) J Chem Soc Chem Commun 1013
63. Batler JH, Spina KP (1984) Synth Commun 14
64. Barrett AGM, Cramp SM, Roberts RS (1999) Org Lett 1:1083
65. Barrett AGM, Hopkins BT, Köbberling J (2002) Chem Rev 102:3301

66. Buchmeiser MR, Atzl N, Bonn GK (1997) J Am Chem Soc 119:9166
67. Silberg J, Schareina T, Kempe R, Wurst K, Buchmeiser MR (2001) J Organomet Chem 622:6
68. Buchmeiser MR, Schareina T, Kempe R, Wurst K (2001) J Organomet Chem 634:39
69. Kröll RM, Schuler N, Lubbad S, Buchmeiser MR (2003) Chem Commun 2742
70. Caster KC, Walls RD (2002) Adv Synth Catal 344:764
71. Klavetter FL, Grubbs RH (1988) J Am Chem Soc 110:7807
72. Afeyan NB, Gordon NF, Mazsaroff I, Varady L, Fulton SP, Yang YB, Regnier FE (1990) J Chromatogr 519:1
73. Afeyan NB, Fulton SP, Regnier FE (1991) J Chromatogr 544:267
74. Kubin M, Spacek P, Chromecek R (1967) Collect Czech Chem Commun 32:3881
75. Hansen LC, Sievers RE (1974) J Chromatogr 99:123
76. Hosoya K, Ohta H, Yoshizoka K, Kimatas K, Ikegami T, Tanaka N (1999) J Chromatogr A 853:11
77. Maruska A, Ericson C, Végvári A, Hjertén S (1999) J Chromatogr A 837:25
78. Gusev I, Huang X, Horváth C (1999) J Chromatogr A 855:273
79. Tang Q, Xin B, Lee ML (1999) J Chromatogr A 837:35
80. Xie S, Svec F, Fréchet JMJ (1998) Chem Mater 10:4072
81. Gerstner JA, Hamilton R, Cramer SM (1992) J Chromatogr 596:173
82. Tanaka N, Nagayama H, Kobayashi H, Ikegami T, Hosoya K, Ishizuka N, Minakuchi H, Nakanishi K, Cabrera K, Lubda D (2000) J High Res Chromatogr 23:111
83. Tanaka N, Kobayashi H, Ishizuka N, Minakuchi H, Nakanishi K, Hosoya K, Ikegami T (2002) J Chromatogr A 965:35
84. Rabel F, Cabrera K, Lubda D (2001) Int Lab 01/02:23
85. Cabrera K, Lubda D, Eggenweiler H-M, Minakuchi H, Nakanishi K (2000) J High Res Chromatogr 23:93
86. Sinner F, Buchmeiser MR (2000) Macromolecules 33:5777
87. Sinner F, Buchmeiser MR (2000) Angew Chem 112:1491
88. Mayr B, Tessadri R, Post E, Buchmeiser MR (2001) Anal Chem 73:4071
89. Mayr M, Mayr B, Buchmeiser MR (2001) Angew Chem 113:3957
90. Mayr B, Hölzl G, Eder K, Buchmeiser MR, Huber CG (2002) Anal Chem 74:6080
91. Lubbad S, Mayr B, Mayr M, Buchmeiser MR (2004) Macromol Symp 210:1
92. Lubbad S, Mayr B, Huber CG, Buchmeiser MR (2002) J Chromatogr A 959:121
93. Lubbad S, Buchmeiser MR (2002) Macromol Rapid Commun 23:617
94. Lubbad S, Buchmeiser MR (2003) Macromol Rapid Commun 24:580
95. Buchmeiser MR, Sinner F (1999) European Pat Appl 409:095 (A 960/99, PCT/EP00/ 04 768)
96. Buchmeiser MR (2001) Macromol Rapid Commun 22:1081
97. Buchmeiser MR (2002) J Mol Catal A: Chemical 190:145
98. Buchmeiser MR (2003) In: Scvec F, Tennikova TB, Deyl Z (eds) Monolithic materials: Preparation, properties and applications (J Chromatogr Library), vol 67. Elsevier, Amsterdam
99. Ertl G, Knözinger H, Weitkamp J (1999) Preparation of solid catalysts. Wiley, Weinheim
100. Peters EC, Svec F, Fréchet JMJ (1999) Adv Mater 11:1169
101. Buchmeiser MR (2001) Angew Chem 113:3911
102. Rodrigues AE (1997) J Chromatogr B 699:47
103. Xu Y, Liapis AI (1996) J Chromatogr A 724:13
104. Sykora D, Svec F, Fréchet JMJ (1999) J Chromatogr A 852:297
105. Viklund C, Pontén E, Glad B, Irgum K, Hörstedt P, Svec F (1997) Chem Mater 9:463

106. Viklund C, Svec F, Fréchet JMJ, Irgum K (1996) Chem Mater 8:744
107. Wang QC, Svec F, Fréchet JMJ (1993) Anal Chem 65:2243
108. Tennikova TB, Deyl Z, Svec F (2001) Metathesis-based monolithic supports: Synthesis, functionalization and applications. Elsevier, Amsterdam
109. Schrock RR (1990) Acc Chem Res 23:158
110. Schrock RR (2001) J Chem Soc Dalton Trans 2541
111. Halász I, Martin K (1975) Ber Bunsen Phys Chem 79:731
112. Halász I, Martin K (1978) Angew Chem 90:954
113. Leon y Leon CA, Thomas MA (1997) GIT Lab J 2:101
114. Szwarc M (1992) Makromol Chem Rapid Commun 13:141
115. Penczek S, Kubisa P, Szymanski R (1991) Makromol Chem Rapid Commun 12:77
116. Johnson AF, Mohsin MA, Meszena ZG, Graves-Morris P (1999) J Macromol Sci Rev Macromol Chem Phys C39:527
117. Szwarc M (1998) J Polym Sci Polym Chem 36:ix
118. Webster OW (1991) Science 251:887
119. Deleuze H, Faivrea R, Herroguez V (2002) Chem Commun 2822
120. Nandakumar MP, Pålsson E, Gustavsson P-E, Larsson P-O, Mattiasson B (2001) Bioseparation 9:193
121. Scholl M, Trnka TM, Morgan JP, Grubbs RH (1999) Tetrahedron Lett 40:2247
122. Scholl M, Ding S, Lee CW, Grubbs RH (1999) Org Lett 1:953
123. Buchmeiser MR (2002) Bioorg Med Chem Lett 12:1837
124. Mayr M, Mayr B, Buchmeiser MR (2002) Des Monomers Polymers 5:325
125. Buchmeiser MR, Mayr M, Mayr B (2001) Austrian Patent Application A 409095 219
126. Buchmeiser MR, Nuyken O, Krause JO (2002) Austrian Patent Application A 411760 (1344/2002)
127. Krause JO, Lubbad S, Mayr M, Nuyken O, Buchmeiser MR (2003) Polym Prepr 44:790
128. Krause JO, Lubbad SH, Nuyken O, Buchmeiser MR (2003) Macromol Rapid Commun 24:875
129. Buchmeiser MR (2004) New J Chem 28:549
130. Buchmeiser MR, Wurst K (1999) J Am Chem Soc 121:11101
131. Gómez FJ, Chen RJ, Wang D, Waymouth RM, Dai H (2003) Chem Commun 190

Author Index Volumes 101–197

Author Index Volumes 1–100 see Volume 100

de Abajo, J. and *de la Campa, J. G.*: Processable Aromatic Polyimides. Vol. 140, pp. 23–60.
Abe, A., Furuya, H., Zhou, Z., Hiejima, T. and *Kobayashi, Y.*: Stepwise Phase Transitions of Chain Molecules: Crystallization/Melting via a Nematic Liquid-Crystalline Phase. Vol. 181, pp. 121–152.
Abetz, V. and *Simon, P. F. W.*: Phase Behaviour and Morphologies of Block Copolymers. Vol. 189, pp. 125–212.
Abetz, V. see Förster, S.: Vol. 166, pp. 173–210.
Adolf, D. B. see Ediger, M. D.: Vol. 116, pp. 73–110.
Advincula R.: Polymer Brushes by Anionic and Cationic Surface-Initiated Polymerization (SIP). Vol. 197, pp. 107–136.
Aharoni, S. M. and *Edwards, S. F.*: Rigid Polymer Networks. Vol. 118, pp. 1–231.
Alakhov, V. Y. see Kabanov, A. V.: Vol. 193, pp. 173–198.
Albertsson, A.-C. and *Varma, I. K.*: Aliphatic Polyesters: Synthesis, Properties and Applications. Vol. 157, pp. 99–138.
Albertsson, A.-C. see Edlund, U.: Vol. 157, pp. 53–98.
Albertsson, A.-C. see Söderqvist Lindblad, M.: Vol. 157, pp. 139–161.
Albertsson, A.-C. see Stridsberg, K. M.: Vol. 157, pp. 27–51.
Albertsson, A.-C. see Al-Malaika, S.: Vol. 169, pp. 177–199.
Allegra, G. and *Meille, S. V.*: Pre-Crystalline, High-Entropy Aggregates: A Role in Polymer Crystallization? Vol. 191, pp. 87–135.
Allen, S. see Ellis, J. S.: Vol. 193, pp. 123–172.
Al-Malaika, S.: Perspectives in Stabilisation of Polyolefins. Vol. 169, pp. 121–150.
Altstädt, V.: The Influence of Molecular Variables on Fatigue Resistance in Stress Cracking Environments. Vol. 188, pp. 105–152.
Améduri, B., Boutevin, B. and *Gramain, P.*: Synthesis of Block Copolymers by Radical Polymerization and Telomerization. Vol. 127, pp. 87–142.
Améduri, B. and *Boutevin, B.*: Synthesis and Properties of Fluorinated Telechelic Monodispersed Compounds. Vol. 102, pp. 133–170.
Ameduri, B. see Taguet, A.: Vol. 184, pp. 127–211.
Amir, R. J. and *Shabat, D.*: Domino Dendrimers. Vol. 192, pp. 59–94.
Amselem, S. see Domb, A. J.: Vol. 107, pp. 93–142.
Anantawaraskul, S., Soares, J. B. P. and *Wood-Adams, P. M.*: Fractionation of Semicrystalline Polymers by Crystallization Analysis Fractionation and Temperature Rising Elution Fractionation. Vol. 182, pp. 1–54.
Andrady, A. L.: Wavelenght Sensitivity in Polymer Photodegradation. Vol. 128, pp. 47–94.
Andreis, M. and *Koenig, J. L.*: Application of Nitrogen–15 NMR to Polymers. Vol. 124, pp. 191–238.
Angiolini, L. see Carlini, C.: Vol. 123, pp. 127–214.
Anjum, N. see Gupta, B.: Vol. 162, pp. 37–63.

Anseth, K. S., Newman, S. M. and *Bowman, C. N.:* Polymeric Dental Composites: Properties and Reaction Behavior of Multimethacrylate Dental Restorations. Vol. 122, pp. 177–218.
Antonietti, M. see Cölfen, H.: Vol. 150, pp. 67–187.
Aoki, H. see Ito, S.: Vol. 182, pp. 131–170.
Armitage, B. A. see O'Brien, D. F.: Vol. 126, pp. 53–58.
Arnal, M. L. see Müller, A. J.: Vol. 190, pp. 1–63.
Arndt, M. see Kaminski, W.: Vol. 127, pp. 143–187.
Arnold, A. and *Holm, C.:* Efficient Methods to Compute Long-Range Interactions for Soft Matter Systems. Vol. 185, pp. 59–109.
Arnold Jr., F. E. and *Arnold, F. E.:* Rigid-Rod Polymers and Molecular Composites. Vol. 117, pp. 257–296.
Arora, M. see Kumar, M. N. V. R.: Vol. 160, pp. 45–118.
Arshady, R.: Polymer Synthesis via Activated Esters: A New Dimension of Creativity in Macromolecular Chemistry. Vol. 111, pp. 1–42.
Aseyev, V. O., Tenhu, H. and *Winnik, F. M.:* Temperature Dependence of the Colloidal Stability of Neutral Amphiphilic Polymers in Water. Vol. 196, pp. 1–86.
Auer, S. and *Frenkel, D.:* Numerical Simulation of Crystal Nucleation in Colloids. Vol. 173, pp. 149–208.
Auriemma, F., de Rosa, C. and *Corradini, P.:* Solid Mesophases in Semicrystalline Polymers: Structural Analysis by Diffraction Techniques. Vol. 181, pp. 1–74.

Bahar, I., Erman, B. and *Monnerie, L.:* Effect of Molecular Structure on Local Chain Dynamics: Analytical Approaches and Computational Methods. Vol. 116, pp. 145–206.
Baietto-Dubourg, M. C. see Chateauminois, A.: Vol. 188, pp. 153–193.
Ballauff, M. see Dingenouts, N.: Vol. 144, pp. 1–48.
Ballauff, M. see Holm, C.: Vol. 166, pp. 1–27.
Ballauff, M. see Rühe, J.: Vol. 165, pp. 79–150.
Balsamo, V. see Müller, A. J.: Vol. 190, pp. 1–63.
Baltá-Calleja, F. J., González Arche, A., Ezquerra, T. A., Santa Cruz, C., Batallón, F., Frick, B. and *López Cabarcos, E.:* Structure and Properties of Ferroelectric Copolymers of Poly(vinylidene) Fluoride. Vol. 108, pp. 1–48.
Baltussen, J. J. M. see Northolt, M. G.: Vol. 178, pp. 1–108.
Barnes, M. D. see Otaigbe, J. U.: Vol. 154, pp. 1–86.
Barnes, C. M. see Satchi-Fainaro, R.: Vol. 193, pp. 1–65.
Barsett, H. see Paulsen, S. B.: Vol. 186, pp. 69–101.
Barshtein, G. R. and *Sabsai, O. Y.:* Compositions with Mineralorganic Fillers. Vol. 101, pp. 1–28.
Barton, J. see Hunkeler, D.: Vol. 112, pp. 115–134.
Baschnagel, J., Binder, K., Doruker, P., Gusev, A. A., Hahn, O., Kremer, K., Mattice, W. L., Müller-Plathe, F., Murat, M., Paul, W., Santos, S., Sutter, U. W. and *Tries, V.:* Bridging the Gap Between Atomistic and Coarse-Grained Models of Polymers: Status and Perspectives. Vol. 152, pp. 41–156.
Bassett, D. C.: On the Role of the Hexagonal Phase in the Crystallization of Polyethylene. Vol. 180, pp. 1–16.
Batallán, F. see Baltá-Calleja, F. J.: Vol. 108, pp. 1–48.
Batog, A. E., Pet'ko, I. P. and *Penczek, P.:* Aliphatic-Cycloaliphatic Epoxy Compounds and Polymers. Vol. 144, pp. 49–114.
Batrakova, E. V. see Kabanov, A. V.: Vol. 193, pp. 173–198.
Baughman, T. W. and *Wagener, K. B.:* Recent Advances in ADMET Polymerization. Vol. 176, pp. 1–42.

Becker, O. and *Simon, G. P.*: Epoxy Layered Silicate Nanocomposites. Vol. 179, pp. 29–82.
Bell, C. L. and *Peppas, N. A.*: Biomedical Membranes from Hydrogels and Interpolymer Complexes. Vol. 122, pp. 125–176.
Bellon-Maurel, A. see Calmon-Decriaud, A.: Vol. 135, pp. 207–226.
Bennett, D. E. see O'Brien, D. F.: Vol. 126, pp. 53–84.
Berry, G. C.: Static and Dynamic Light Scattering on Moderately Concentraded Solutions: Isotropic Solutions of Flexible and Rodlike Chains and Nematic Solutions of Rodlike Chains. Vol. 114, pp. 233–290.
Bershtein, V. A. and *Ryzhov, V. A.*: Far Infrared Spectroscopy of Polymers. Vol. 114, pp. 43–122.
Bhargava, R., Wang, S.-Q. and *Koenig, J. L*: FTIR Microspectroscopy of Polymeric Systems. Vol. 163, pp. 137–191.
Biesalski, M. see Rühe, J.: Vol. 165, pp. 79–150.
Bigg, D. M.: Thermal Conductivity of Heterophase Polymer Compositions. Vol. 119, pp. 1–30.
Binder, K.: Phase Transitions in Polymer Blends and Block Copolymer Melts: Some Recent Developments. Vol. 112, pp. 115–134.
Binder, K.: Phase Transitions of Polymer Blends and Block Copolymer Melts in Thin Films. Vol. 138, pp. 1–90.
Binder, K. see Baschnagel, J.: Vol. 152, pp. 41–156.
Binder, K., Müller, M., Virnau, P. and *González MacDowell, L.*: Polymer+Solvent Systems: Phase Diagrams, Interface Free Energies, and Nucleation. Vol. 173, pp. 1–104.
Bird, R. B. see Curtiss, C. F.: Vol. 125, pp. 1–102.
Biswas, M. and *Mukherjee, A.*: Synthesis and Evaluation of Metal-Containing Polymers. Vol. 115, pp. 89–124.
Biswas, M. and *Sinha Ray, S.*: Recent Progress in Synthesis and Evaluation of Polymer-Montmorillonite Nanocomposites. Vol. 155, pp. 167–221.
Blankenburg, L. see Klemm, E.: Vol. 177, pp. 53–90.
Blumen, A. see Gurtovenko, A. A.: Vol. 182, pp. 171–282.
Bogdal, D., Penczek, P., Pielichowski, J. and *Prociak, A.*: Microwave Assisted Synthesis, Crosslinking, and Processing of Polymeric Materials. Vol. 163, pp. 193–263.
Bohrisch, J., Eisenbach, C. D., Jaeger, W., Mori, H., Müller, A. H. E., Rehahn, M., Schaller, C., Traser, S. and *Wittmeyer, P.*: New Polyelectrolyte Architectures. Vol. 165, pp. 1–41.
Bolze, J. see Dingenouts, N.: Vol. 144, pp. 1–48.
Bosshard, C.: see Gubler, U.: Vol. 158, pp. 123–190.
Boutevin, B. and *Robin, J. J.*: Synthesis and Properties of Fluorinated Diols. Vol. 102, pp. 105–132.
Boutevin, B. see Améduri, B.: Vol. 102, pp. 133–170.
Boutevin, B. see Améduri, B.: Vol. 127, pp. 87–142.
Boutevin, B. see Guida-Pietrasanta, F.: Vol. 179, pp. 1–27.
Boutevin, B. see Taguet, A.: Vol. 184, pp. 127–211.
Bowman, C. N. see Anseth, K. S.: Vol. 122, pp. 177–218.
Boyd, R. H.: Prediction of Polymer Crystal Structures and Properties. Vol. 116, pp. 1–26.
Bracco, S. see Sozzani, P.: Vol. 181, pp. 153–177.
Briber, R. M. see Hedrick, J. L.: Vol. 141, pp. 1–44.
Bronnikov, S. V., Vettegren, V. I. and *Frenkel, S. Y.*: Kinetics of Deformation and Relaxation in Highly Oriented Polymers. Vol. 125, pp. 103–146.
Brown, H. R. see Creton, C.: Vol. 156, pp. 53–135.
Bruza, K. J. see Kirchhoff, R. A.: Vol. 117, pp. 1–66.
Buchmeiser M. R.: Metathesis Polymerization To and From Surfaces. Vol. 197, pp. 137–171.

Buchmeiser, M. R.: Regioselective Polymerization of 1-Alkynes and Stereoselective Cyclopolymerization of a, w-Heptadiynes. Vol. 176, pp. 89–119.
Budkowski, A.: Interfacial Phenomena in Thin Polymer Films: Phase Coexistence and Segregation. Vol. 148, pp. 1–112.
Bunz, U. H. F.: Synthesis and Structure of PAEs. Vol. 177, pp. 1–52.
Burban, J. H. see *Cussler, E. L.*: Vol. 110, pp. 67–80.
Burchard, W.: Solution Properties of Branched Macromolecules. Vol. 143, pp. 113–194.
Butté, A. see *Schork, F. J.*: Vol. 175, pp. 129–255.

Calmon-Decriaud, A., Bellon-Maurel, V., Silvestre, F.: Standard Methods for Testing the Aerobic Biodegradation of Polymeric Materials. Vol. 135, pp. 207–226.
Cameron, N. R. and *Sherrington, D. C.*: High Internal Phase Emulsions (HIPEs)-Structure, Properties and Use in Polymer Preparation. Vol. 126, pp. 163–214.
de la Campa, J. G. see *de Abajo, J.*: Vol. 140, pp. 23–60.
Candau, F. see *Hunkeler, D.*: Vol. 112, pp. 115–134.
Canelas, D. A. and *DeSimone, J. M.*: Polymerizations in Liquid and Supercritical Carbon Dioxide. Vol. 133, pp. 103–140.
Canva, M. and *Stegeman, G. I.*: Quadratic Parametric Interactions in Organic Waveguides. Vol. 158, pp. 87–121.
Capek, I.: Kinetics of the Free-Radical Emulsion Polymerization of Vinyl Chloride. Vol. 120, pp. 135–206.
Capek, I.: Radical Polymerization of Polyoxyethylene Macromonomers in Disperse Systems. Vol. 145, pp. 1–56.
Capek, I. and *Chern, C.-S.*: Radical Polymerization in Direct Mini-Emulsion Systems. Vol. 155, pp. 101–166.
Cappella, B. see *Munz, M.*: Vol. 164, pp. 87–210.
Carlesso, G. see *Prokop, A.*: Vol. 160, pp. 119–174.
Carlini, C. and *Angiolini, L.*: Polymers as Free Radical Photoinitiators. Vol. 123, pp. 127–214.
Carter, K. R. see *Hedrick, J. L.*: Vol. 141, pp. 1–44.
Casas-Vazquez, J. see *Jou, D.*: Vol. 120, pp. 207–266.
Chan, C.-M. and *Li, L.*: Direct Observation of the Growth of Lamellae and Spherulites by AFM. Vol. 188, pp. 1–41.
Chandrasekhar, V.: Polymer Solid Electrolytes: Synthesis and Structure. Vol. 135, pp. 139–206.
Chang, J. Y. see *Han, M. J.*: Vol. 153, pp. 1–36.
Chang, T.: Recent Advances in Liquid Chromatography Analysis of Synthetic Polymers. Vol. 163, pp. 1–60.
Charleux, B. and *Faust, R.*: Synthesis of Branched Polymers by Cationic Polymerization. Vol. 142, pp. 1–70.
Chateauminois, A. and *Baietto-Dubourg, M. C.*: Fracture of Glassy Polymers Within Sliding Contacts. Vol. 188, pp. 153–193.
Chen, P. see *Jaffe, M.*: Vol. 117, pp. 297–328.
Chern, C.-S. see *Capek, I.*: Vol. 155, pp. 101–166.
Chevolot, Y. see *Mathieu, H. J.*: Vol. 162, pp. 1–35.
Chim, Y. T. A. see *Ellis, J. S.*: Vol. 193, pp. 123–172.
Choe, E.-W. see *Jaffe, M.*: Vol. 117, pp. 297–328.
Chow, P. Y. and *Gan, L. M.*: Microemulsion Polymerizations and Reactions. Vol. 175, pp. 257–298.
Chow, T. S.: Glassy State Relaxation and Deformation in Polymers. Vol. 103, pp. 149–190.
Chujo, Y. see *Uemura, T.*: Vol. 167, pp. 81–106.

Chung, S.-J. see Lin, T.-C.: Vol. 161, pp. 157–193.
Chung, T.-S. see Jaffe, M.: Vol. 117, pp. 297–328.
Clarke, N.: Effect of Shear Flow on Polymer Blends. Vol. 183, pp. 127–173.
Coenjarts, C. see Li, M.: Vol. 190, pp. 183–226.
Cölfen, H. and *Antonietti, M.*: Field-Flow Fractionation Techniques for Polymer and Colloid Analysis. Vol. 150, pp. 67–187.
Colmenero, J. see Richter, D.: Vol. 174, pp. 1–221.
Comanita, B. see Roovers, J.: Vol. 142, pp. 179–228.
Comotti, A. see Sozzani, P.: Vol. 181, pp. 153–177.
Connell, J. W. see Hergenrother, P. M.: Vol. 117, pp. 67–110.
Corradini, P. see Auriemma, F.: Vol. 181, pp. 1–74.
Creton, C., Kramer, E. J., Brown, H. R. and *Hui, C.-Y.*: Adhesion and Fracture of Interfaces Between Immiscible Polymers: From the Molecular to the Continuum Scale. Vol. 156, pp. 53–135.
Criado-Sancho, M. see Jou, D.: Vol. 120, pp. 207–266.
Curro, J. G. see Schweizer, K. S.: Vol. 116, pp. 319–378.
Curtiss, C. F. and *Bird, R. B.*: Statistical Mechanics of Transport Phenomena: Polymeric Liquid Mixtures. Vol. 125, pp. 1–102.
Cussler, E. L., Wang, K. L. and *Burban, J. H.*: Hydrogels as Separation Agents. Vol. 110, pp. 67–80.
Czub, P. see Penczek, P.: Vol. 184, pp. 1–95.

Dalton, L.: Nonlinear Optical Polymeric Materials: From Chromophore Design to Commercial Applications. Vol. 158, pp. 1–86.
Dautzenberg, H. see Holm, C.: Vol. 166, pp. 113–171.
Davidson, J. M. see Prokop, A.: Vol. 160, pp. 119–174.
Davies, M. C. see Ellis, J. S.: Vol. 193, pp. 123–172.
Den Decker, M. G. see Northolt, M. G.: Vol. 178, pp. 1–108.
Desai, S. M. and *Singh, R. P.*: Surface Modification of Polyethylene. Vol. 169, pp. 231–293.
DeSimone, J. M. see Canelas, D. A.: Vol. 133, pp. 103–140.
DeSimone, J. M. see Kennedy, K. A.: Vol. 175, pp. 329–346.
Dhal, P. K., Holmes-Farley, S. R., Huval, C. C. and *Jozefiak, T. H.*: Polymers as Drugs. Vol. 192, pp. 9–58.
DiMari, S. see Prokop, A.: Vol. 136, pp. 1–52.
Dimonie, M. V. see Hunkeler, D.: Vol. 112, pp. 115–134.
Dingenouts, N., Bolze, J., Pötschke, D. and *Ballauf, M.*: Analysis of Polymer Latexes by Small-Angle X-Ray Scattering. Vol. 144, pp. 1–48.
Dodd, L. R. and *Theodorou, D. N.*: Atomistic Monte Carlo Simulation and Continuum Mean Field Theory of the Structure and Equation of State Properties of Alkane and Polymer Melts. Vol. 116, pp. 249–282.
Doelker, E.: Cellulose Derivatives. Vol. 107, pp. 199–266.
Dolden, J. G.: Calculation of a Mesogenic Index with Emphasis Upon LC-Polyimides. Vol. 141, pp. 189–245.
Domb, A. J., Amselem, S., Shah, J. and *Maniar, M.*: Polyanhydrides: Synthesis and Characterization. Vol. 107, pp. 93–142.
Domb, A. J. see Kumar, M. N. V. R.: Vol. 160, pp. 45–118.
Doruker, P. see Baschnagel, J.: Vol. 152, pp. 41–156.
Dubois, P. see Mecerreyes, D.: Vol. 147, pp. 1–60.
Dubrovskii, S. A. see Kazanskii, K. S.: Vol. 104, pp. 97–134.
Dudowicz, J. see Freed, K. F.: Vol. 183, pp. 63–126.

Duncan, R., Ringsdorf, H. and *Satchi-Fainaro, R.*: Polymer Therapeutics: Polymers as Drugs, Drug and Protein Conjugates and Gene Delivery Systems: Past, Present and Future Opportunities. Vol. 192, pp. 1–8.
Duncan, R. see Satchi-Fainaro, R.: Vol. 193, pp. 1–65.
Dunkin, I. R. see Steinke, J.: Vol. 123, pp. 81–126.
Dunson, D. L. see McGrath, J. E.: Vol. 140, pp. 61–106.
Dyer D. J.: Photoinitiated Synthesis of Grafted Polymers. Vol. 197, pp. 47–65.
Dziezok, P. see Rühe, J.: Vol. 165, pp. 79–150.

Eastmond, G. C.: Poly(e-caprolactone) Blends. Vol. 149, pp. 59–223.
Ebringerová, A., Hromádková, Z. and *Heinze, T.*: Hemicellulose. Vol. 186, pp. 1–67.
Economy, J. and *Goranov, K.*: Thermotropic Liquid Crystalline Polymers for High Performance Applications. Vol. 117, pp. 221–256.
Ediger, M. D. and *Adolf, D. B.*: Brownian Dynamics Simulations of Local Polymer Dynamics. Vol. 116, pp. 73–110.
Edlund, U. and *Albertsson, A.-C.*: Degradable Polymer Microspheres for Controlled Drug Delivery. Vol. 157, pp. 53–98.
Edwards, S. F. see Aharoni, S. M.: Vol. 118, pp. 1–231.
Eisenbach, C. D. see Bohrisch, J.: Vol. 165, pp. 1–41.
Ellis, J. S., Allen, S., Chim, Y. T. A., Roberts, C. J., Tendler, S. J. B. and *Davies, M. C.*: Molecular-Scale Studies on Biopolymers Using Atomic Force Microscopy. Vol. 193, pp. 123–172.
Endo, T. see Yagci, Y.: Vol. 127, pp. 59–86.
Engelhardt, H. and *Grosche, O.*: Capillary Electrophoresis in Polymer Analysis. Vol. 150, pp. 189–217.
Engelhardt, H. and *Martin, H.*: Characterization of Synthetic Polyelectrolytes by Capillary Electrophoretic Methods. Vol. 165, pp. 211–247.
Eriksson, P. see Jacobson, K.: Vol. 169, pp. 151–176.
Erman, B. see Bahar, I.: Vol. 116, pp. 145–206.
Eschner, M. see Spange, S.: Vol. 165, pp. 43–78.
Estel, K. see Spange, S.: Vol. 165, pp. 43–78.
Estevez, R. and *Van der Giessen, E.*: Modeling and Computational Analysis of Fracture of Glassy Polymers. Vol. 188, pp. 195–234.
Ewen, B. and *Richter, D.*: Neutron Spin Echo Investigations on the Segmental Dynamics of Polymers in Melts, Networks and Solutions. Vol. 134, pp. 1–130.
Ezquerra, T. A. see Baltá-Calleja, F. J.: Vol. 108, pp. 1–48.

Fatkullin, N. see Kimmich, R.: Vol. 170, pp. 1–113.
Faust, R. see Charleux, B.: Vol. 142, pp. 1–70.
Faust, R. see Kwon, Y.: Vol. 167, pp. 107–135.
Fekete, E. see Pukánszky, B.: Vol. 139, pp. 109–154.
Fendler, J. H.: Membrane-Mimetic Approach to Advanced Materials. Vol. 113, pp. 1–209.
Fetters, L. J. see Xu, Z.: Vol. 120, pp. 1–50.
Fontenot, K. see Schork, F. J.: Vol. 175, pp. 129–255.
Förster, S., Abetz, V. and *Müller, A. H. E.*: Polyelectrolyte Block Copolymer Micelles. Vol. 166, pp. 173–210.
Förster, S. and *Schmidt, M.*: Polyelectrolytes in Solution. Vol. 120, pp. 51–134.
Freed, K. F. and *Dudowicz, J.*: Influence of Monomer Molecular Structure on the Miscibility of Polymer Blends. Vol. 183, pp. 63–126.
Freire, J. J.: Conformational Properties of Branched Polymers: Theory and Simulations. Vol. 143, pp. 35–112.

Frenkel, D. see Hu, W.: Vol. 191, pp. 1–35.
Frenkel, S. Y. see Bronnikov, S. V.: Vol. 125, pp. 103–146.
Frick, B. see Baltá-Calleja, F. J.: Vol. 108, pp. 1–48.
Fridman, M. L.: see Terent'eva, J. P.: Vol. 101, pp. 29–64.
Fuchs, G. see Trimmel, G.: Vol. 176, pp. 43–87.
Fukuda, T. see Tsujii, Y.: Vol. 197, pp. 1–47.
Fukui, K. see Otaigbe, J. U.: Vol. 154, pp. 1–86.
Funke, W.: Microgels-Intramolecularly Crosslinked Macromolecules with a Globular Structure. Vol. 136, pp. 137–232.
Furusho, Y. see Takata, T.: Vol. 171, pp. 1–75.
Furuya, H. see Abe, A.: Vol. 181, pp. 121–152.

Galina, H.: Mean-Field Kinetic Modeling of Polymerization: The Smoluchowski Coagulation Equation. Vol. 137, pp. 135–172.
Gan, L. M. see Chow, P. Y.: Vol. 175, pp. 257–298.
Ganesh, K. see Kishore, K.: Vol. 121, pp. 81–122.
Gaw, K. O. and *Kakimoto, M.*: Polyimide-Epoxy Composites. Vol. 140, pp. 107–136.
Geckeler, K. E. see Rivas, B.: Vol. 102, pp. 171–188.
Geckeler, K. E.: Soluble Polymer Supports for Liquid-Phase Synthesis. Vol. 121, pp. 31–80.
Gedde, U. W. and *Mattozzi, A.*: Polyethylene Morphology. Vol. 169, pp. 29–73.
Gehrke, S. H.: Synthesis, Equilibrium Swelling, Kinetics Permeability and Applications of Environmentally Responsive Gels. Vol. 110, pp. 81–144.
Geil, P. H., Yang, J., Williams, R. A., Petersen, K. L., Long, T.-C. and *Xu, P.*: Effect of Molecular Weight and Melt Time and Temperature on the Morphology of Poly(tetrafluorethylene). Vol. 180, pp. 89–159.
de Gennes, P.-G.: Flexible Polymers in Nanopores. Vol. 138, pp. 91–106.
Georgiou, S.: Laser Cleaning Methodologies of Polymer Substrates. Vol. 168, pp. 1–49.
Geuss, M. see Munz, M.: Vol. 164, pp. 87–210.
Giannelis, E. P., Krishnamoorti, R. and *Manias, E.*: Polymer-Silicate Nanocomposites: Model Systems for Confined Polymers and Polymer Brushes. Vol. 138, pp. 107–148.
Van der Giessen, E. see Estevez, R.: Vol. 188, pp. 195–234.
Godovsky, D. Y.: Device Applications of Polymer-Nanocomposites. Vol. 153, pp. 163–205.
Godovsky, D. Y.: Electron Behavior and Magnetic Properties Polymer-Nanocomposites. Vol. 119, pp. 79–122.
Gohy, J.-F.: Block Copolymer Micelles. Vol. 190, pp. 65–136.
González Arche, A. see Baltá-Calleja, F. J.: Vol. 108, pp. 1–48.
Goranov, K. see Economy, J.: Vol. 117, pp. 221–256.
Goto, A. see Tsujii, Y.: Vol. 197, pp. 1–47.
Gramain, P. see Améduri, B.: Vol. 127, pp. 87–142.
Grein, C.: Toughness of Neat, Rubber Modified and Filled β-Nucleated Polypropylene: From Fundamentals to Applications. Vol. 188, pp. 43–104.
Greish, K. see Maeda, H.: Vol. 193, pp. 103–121.
Grest, G. S.: Normal and Shear Forces Between Polymer Brushes. Vol. 138, pp. 149–184.
Grigorescu, G. and *Kulicke, W.-M.*: Prediction of Viscoelastic Properties and Shear Stability of Polymers in Solution. Vol. 152, p. 1–40.
Gröhn, F. see Rühe, J.: Vol. 165, pp. 79–150.
Grosberg, A. Y. and *Khokhlov, A. R.*: After-Action of the Ideas of I. M. Lifshitz in Polymer and Biopolymer Physics. Vol. 196, pp. 189–210.
Grosberg, A. and *Nechaev, S.*: Polymer Topology. Vol. 106, pp. 1–30.
Grosche, O. see Engelhardt, H.: Vol. 150, pp. 189–217.

Grubbs, R., Risse, W. and *Novac, B.*: The Development of Well-defined Catalysts for Ring-Opening Olefin Metathesis. Vol. 102, pp. 47–72.

Gubler, U. and *Bosshard, C.*: Molecular Design for Third-Order Nonlinear Optics. Vol. 158, pp. 123–190.

Guida-Pietrasanta, F. and *Boutevin, B.*: Polysilalkylene or Silarylene Siloxanes Said Hybrid Silicones. Vol. 179, pp. 1–27.

van *Gunsteren, W. F.* see Gusev, A. A.: Vol. 116, pp. 207–248.

Gupta, B. and *Anjum, N.*: Plasma and Radiation-Induced Graft Modification of Polymers for Biomedical Applications. Vol. 162, pp. 37–63.

Gurtovenko, A. A. and *Blumen, A.*: Generalized Gaussian Structures: Models for Polymer Systems with Complex Topologies. Vol. 182, pp. 171–282.

Gusev, A. A., Müller-Plathe, F., van Gunsteren, W. F. and *Suter, U. W.*: Dynamics of Small Molecules in Bulk Polymers. Vol. 116, pp. 207–248.

Gusev, A. A. see Baschnagel, J.: Vol. 152, pp. 41–156.

Guillot, J. see Hunkeler, D.: Vol. 112, pp. 115–134.

Guyot, A. and *Tauer, K.*: Reactive Surfactants in Emulsion Polymerization. Vol. 111, pp. 43–66.

Hadjichristidis, N., Pispas, S., Pitsikalis, M., Iatrou, H. and *Vlahos, C.*: Asymmetric Star Polymers Synthesis and Properties. Vol. 142, pp. 71–128.

Hadjichristidis, N., Pitsikalis, M. and *Iatrou, H.*: Synthesis of Block Copolymers. Vol. 189, pp. 1–124.

Hadjichristidis, N. see Xu, Z.: Vol. 120, pp. 1–50.

Hadjichristidis, N. see Pitsikalis, M.: Vol. 135, pp. 1–138.

Hahn, O. see Baschnagel, J.: Vol. 152, pp. 41–156.

Hakkarainen, M.: Aliphatic Polyesters: Abiotic and Biotic Degradation and Degradation Products. Vol. 157, pp. 1–26.

Hakkarainen, M. and *Albertsson, A.-C.*: Environmental Degradation of Polyethylene. Vol. 169, pp. 177–199.

Halary, J. L. see Monnerie, L.: Vol. 187, pp. 35–213.

Halary, J. L. see Monnerie, L.: Vol. 187, pp. 215–364.

Hall, H. K. see Penelle, J.: Vol. 102, pp. 73–104.

Hamley, I. W.: Crystallization in Block Copolymers. Vol. 148, pp. 113–138.

Hammouda, B.: SANS from Homogeneous Polymer Mixtures: A Unified Overview. Vol. 106, pp. 87–134.

Han, M. J. and *Chang, J. Y.*: Polynucleotide Analogues. Vol. 153, pp. 1–36.

Harada, A.: Design and Construction of Supramolecular Architectures Consisting of Cyclodextrins and Polymers. Vol. 133, pp. 141–192.

Haralson, M. A. see Prokop, A.: Vol. 136, pp. 1–52.

Harding, S. E.: Analysis of Polysaccharides by Ultracentrifugation. Size, Conformation and Interactions in Solution. Vol. 186, pp. 211–254.

Hasegawa, N. see Usuki, A.: Vol. 179, pp. 135–195.

Hassan, C. M. and *Peppas, N. A.*: Structure and Applications of Poly(vinyl alcohol) Hydrogels Produced by Conventional Crosslinking or by Freezing/Thawing Methods. Vol. 153, pp. 37–65.

Hawker, C. J.: Dentritic and Hyperbranched Macromolecules Precisely Controlled Macromolecular Architectures. Vol. 147, pp. 113–160.

Hawker, C. J. see Hedrick, J. L.: Vol. 141, pp. 1–44.

He, G. S. see Lin, T.-C.: Vol. 161, pp. 157–193.

Hedrick, J. L., Carter, K. R., Labadie, J. W., Miller, R. D., Volksen, W., Hawker, C. J., Yoon, D. Y., Russell, T. P., McGrath, J. E. and *Briber, R. M.*: Nanoporous Polyimides. Vol. 141, pp. 1–44.

Hedrick, J. L., Labadie, J. W., Volksen, W. and *Hilborn, J. G.*: Nanoscopically Engineered Polyimides. Vol. 147, pp. 61–112.
Hedrick, J. L. see Hergenrother, P. M.: Vol. 117, pp. 67–110.
Hedrick, J. L. see Kiefer, J.: Vol. 147, pp. 161–247.
Hedrick, J. L. see McGrath, J. E.: Vol. 140, pp. 61–106.
Heine, D. R., Grest, G. S. and *Curro, J. G.*: Structure of Polymer Melts and Blends: Comparison of Integral Equation theory and Computer Sumulation. Vol. 173, pp. 209–249.
Heinrich, G. and *Klüppel, M.*: Recent Advances in the Theory of Filler Networking in Elastomers. Vol. 160, pp. 1–44.
Heinze, T. see Ebringerová, A.: Vol. 186, pp. 1–67.
Heinze, T. see El Seoud, O. A.: Vol. 186, pp. 103–149.
Heller, J.: Poly (Ortho Esters). Vol. 107, pp. 41–92.
Helm, C. A. see Möhwald, H.: Vol. 165, pp. 151–175.
Hemielec, A. A. see Hunkeler, D.: Vol. 112, pp. 115–134.
Hergenrother, P. M., Connell, J. W., Labadie, J. W. and *Hedrick, J. L.*: Poly(arylene ether)s Containing Heterocyclic Units. Vol. 117, pp. 67–110.
Hernández-Barajas, J. see Wandrey, C.: Vol. 145, pp. 123–182.
Hervet, H. see Léger, L.: Vol. 138, pp. 185–226.
Hiejima, T. see Abe, A.: Vol. 181, pp. 121–152.
Hikosaka, M., Watanabe, K., Okada, K. and *Yamazaki, S.*: Topological Mechanism of Polymer Nucleation and Growth – The Role of Chain Sliding Diffusion and Entanglement. Vol. 191, pp. 137–186.
Hilborn, J. G. see Hedrick, J. L.: Vol. 147, pp. 61–112.
Hilborn, J. G. see Kiefer, J.: Vol. 147, pp. 161–247.
Hillborg, H. see Vancso, G. J.: Vol. 182, pp. 55–129.
Hillmyer, M. A.: Nanoporous Materials from Block Copolymer Precursors. Vol. 190, pp. 137–181.
Hiramatsu, N. see Matsushige, M.: Vol. 125, pp. 147–186.
Hirasa, O. see Suzuki, M.: Vol. 110, pp. 241–262.
Hirotsu, S.: Coexistence of Phases and the Nature of First-Order Transition in Poly-N-isopropylacrylamide Gels. Vol. 110, pp. 1–26.
Höcker, H. see Klee, D.: Vol. 149, pp. 1–57.
Holm, C. see Arnold, A.: Vol. 185, pp. 59–109.
Holm, C., Hofmann, T., Joanny, J. F., Kremer, K., Netz, R. R., Reineker, P., Seidel, C., Vilgis, T. A. and *Winkler, R. G.*: Polyelectrolyte Theory. Vol. 166, pp. 67–111.
Holm, C., Rehahn, M., Oppermann, W. and *Ballauff, M.*: Stiff-Chain Polyelectrolytes. Vol. 166, pp. 1–27.
Holmes-Farley, S. R. see Dhal, P. K.: Vol. 192, pp. 9–58.
Hornsby, P.: Rheology, Compounding and Processing of Filled Thermoplastics. Vol. 139, pp. 155–216.
Houbenov, N. see Rühe, J.: Vol. 165, pp. 79–150.
Hromádková, Z. see Ebringerová, A.: Vol. 186, pp. 1–67.
Hu, W. and *Frenkel, D.*: Polymer Crystallization Driven by Anisotropic Interactions. Vol. 191, pp. 1–35.
Huber, K. see Volk, N.: Vol. 166, pp. 29–65.
Hugenberg, N. see Rühe, J.: Vol. 165, pp. 79–150.
Hui, C.-Y. see Creton, C.: Vol. 156, pp. 53–135.
Hult, A., Johansson, M. and *Malmström, E.*: Hyperbranched Polymers. Vol. 143, pp. 1–34.
Hünenberger, P. H.: Thermostat Algorithms for Molecular-Dynamics Simulations. Vol. 173, pp. 105–147.

Hunkeler, D., Candau, F., Pichot, C., Hemielec, A. E., Xie, T. Y., Barton, J., Vaskova, V., Guillot, J., Dimonie, M. V. and *Reichert, K. H.*: Heterophase Polymerization: A Physical and Kinetic Comparision and Categorization. Vol. 112, pp. 115–134.
Hunkeler, D. see Macko, T.: Vol. 163, pp. 61–136.
Hunkeler, D. see Prokop, A.: Vol. 136, pp. 1–52; 53–74.
Hunkeler, D. see Wandrey, C.: Vol. 145, pp. 123–182.
Huval, C. C. see Dhal, P. K.: Vol. 192, pp. 9–58.

Iatrou, H. see Hadjichristidis, N.: Vol. 142, pp. 71–128.
Iatrou, H. see Hadjichristidis, N.: Vol. 189, pp. 1–124.
Ichikawa, T. see Yoshida, H.: Vol. 105, pp. 3–36.
Ihara, E. see Yasuda, H.: Vol. 133, pp. 53–102.
Ikada, Y. see Uyama, Y.: Vol. 137, pp. 1–40.
Ikehara, T. see Jinnuai, H.: Vol. 170, pp. 115–167.
Ilavsky, M.: Effect on Phase Transition on Swelling and Mechanical Behavior of Synthetic Hydrogels. Vol. 109, pp. 173–206.
Imai, M. see Kaji, K.: Vol. 191, pp. 187–240.
Imai, Y.: Rapid Synthesis of Polyimides from Nylon-Salt Monomers. Vol. 140, pp. 1–23.
Inomata, H. see Saito, S.: Vol. 106, pp. 207–232.
Inoue, S. see Sugimoto, H.: Vol. 146, pp. 39–120.
Irie, M.: Stimuli-Responsive Poly(N-isopropylacrylamide), Photo- and Chemical-Induced Phase Transitions. Vol. 110, pp. 49–66.
Ise, N. see Matsuoka, H.: Vol. 114, pp. 187–232.
Ishikawa, T.: Advances in Inorganic Fibers. Vol. 178, pp. 109–144.
Ito, H.: Chemical Amplification Resists for Microlithography. Vol. 172, pp. 37–245.
Ito, K. and *Kawaguchi, S.*: Poly(macronomers), Homo- and Copolymerization. Vol. 142, pp. 129–178.
Ito, K. see Kawaguchi, S.: Vol. 175, pp. 299–328.
Ito, S. and *Aoki, H.*: Nano-Imaging of Polymers by Optical Microscopy. Vol. 182, pp. 131–170.
Ito, Y. see Suginome, M.: Vol. 171, pp. 77–136.
Ivanov, A. E. see Zubov, V. P.: Vol. 104, pp. 135–176.

Jacob, S. and *Kennedy, J.*: Synthesis, Characterization and Properties of OCTA-ARM Polyisobutylene-Based Star Polymers. Vol. 146, pp. 1–38.
Jacobson, K., Eriksson, P., Reitberger, T. and *Stenberg, B.*: Chemiluminescence as a Tool for Polyolefin. Vol. 169, pp. 151–176.
Jaeger, W. see Bohrisch, J.: Vol. 165, pp. 1–41.
Jaffe, M., Chen, P., Choe, E.-W., Chung, T.-S. and *Makhija, S.*: High Performance Polymer Blends. Vol. 117, pp. 297–328.
Jancar, J.: Structure-Property Relationships in Thermoplastic Matrices. Vol. 139, pp. 1–66.
Jen, A. K.-Y. see Kajzar, F.: Vol. 161, pp. 1–85.
Jerome, R. see Mecerreyes, D.: Vol. 147, pp. 1–60.
de Jeu, W. H. see Li, L.: Vol. 181, pp. 75–120.
Jiang, M., Li, M., Xiang, M. and *Zhou, H.*: Interpolymer Complexation and Miscibility and Enhancement by Hydrogen Bonding. Vol. 146, pp. 121–194.
Jin, J. see Shim, H.-K.: Vol. 158, pp. 191–241.
Jinnai, H., Nishikawa, Y., Ikehara, T. and *Nishi, T.*: Emerging Technologies for the 3D Analysis of Polymer Structures. Vol. 170, pp. 115–167.
Jo, W. H. and *Yang, J. S.*: Molecular Simulation Approaches for Multiphase Polymer Systems. Vol. 156, pp. 1–52.

Joanny, J.-F. see Holm, C.: Vol. 166, pp. 67–111.
Joanny, J.-F. see Thünemann, A. F.: Vol. 166, pp. 113–171.
Johannsmann, D. see Rühe, J.: Vol. 165, pp. 79–150.
Johansson, M. see Hult, A.: Vol. 143, pp. 1–34.
Joos-Müller, B. see Funke, W.: Vol. 136, pp. 137–232.
Jou, D., Casas-Vazquez, J. and *Criado-Sancho, M.*: Thermodynamics of Polymer Solutions under Flow: Phase Separation and Polymer Degradation. Vol. 120, pp. 207–266.
Jozefiak, T. H. see Dhal, P. K.: Vol. 192, pp. 9–58.

Kabanov, A. V., Batrakova, E. V., Sherman, S. and *Alakhov, V. Y.*: Polymer Genomics. Vol. 193, pp. 173–198.
Kaetsu, I.: Radiation Synthesis of Polymeric Materials for Biomedical and Biochemical Applications. Vol. 105, pp. 81–98.
Kaji, K., Nishida, K., Kanaya, T., Matsuba, G., Konishi, T. and *Imai, M.*: Spinodal Crystallization of Polymers: Crystallization from the Unstable Melt. Vol. 191, pp. 187–240.
Kaji, K. see Kanaya, T.: Vol. 154, pp. 87–141.
Kajzar, F., Lee, K.-S. and *Jen, A. K.-Y.*: Polymeric Materials and their Orientation Techniques for Second-Order Nonlinear Optics. Vol. 161, pp. 1–85.
Kakimoto, M. see Gaw, K. O.: Vol. 140, pp. 107–136.
Kaminski, W. and *Arndt, M.*: Metallocenes for Polymer Catalysis. Vol. 127, pp. 143–187.
Kammer, H. W., Kressler, H. and *Kummerloewe, C.*: Phase Behavior of Polymer Blends – Effects of Thermodynamics and Rheology. Vol. 106, pp. 31–86.
Kanaya, T. and *Kaji, K.*: Dynamcis in the Glassy State and Near the Glass Transition of Amorphous Polymers as Studied by Neutron Scattering. Vol. 154, pp. 87–141.
Kanaya, T. see Kaji, K.: Vol. 191, pp. 187–240.
Kandyrin, L. B. and *Kuleznev, V. N.*: The Dependence of Viscosity on the Composition of Concentrated Dispersions and the Free Volume Concept of Disperse Systems. Vol. 103, pp. 103–148.
Kaneko, M. see Ramaraj, R.: Vol. 123, pp. 215–242.
Kang, E. T., Neoh, K. G. and *Tan, K. L.*: X-Ray Photoelectron Spectroscopic Studies of Electroactive Polymers. Vol. 106, pp. 135–190.
Kaplan, D. L. see Singh, A.: Vol. 194, pp. 211–224.
Kaplan, D. L. see Xu, P.: Vol. 194, pp. 69–94.
Karlsson, S. see Söderqvist Lindblad, M.: Vol. 157, pp. 139–161.
Karlsson, S.: Recycled Polyolefins. Material Properties and Means for Quality Determination. Vol. 169, pp. 201–229.
Kataoka, K. see Nishiyama, N.: Vol. 193, pp. 67–101.
Kato, K. see Uyama, Y.: Vol. 137, pp. 1–40.
Kato, M. see Usuki, A.: Vol. 179, pp. 135–195.
Kausch, H.-H. and *Michler, G. H.*: The Effect of Time on Crazing and Fracture. Vol. 187, pp. 1–33.
Kausch, H.-H. see Monnerie, L. Vol. 187, pp. 215–364.
Kautek, W. see Krüger, J.: Vol. 168, pp. 247–290.
Kawaguchi, S. see Ito, K.: Vol. 142, pp. 129–178.
Kawaguchi, S. and *Ito, K.*: Dispersion Polymerization. Vol. 175, pp. 299–328.
Kawata, S. see Sun, H.-B.: Vol. 170, pp. 169–273.
Kazanskii, K. S. and *Dubrovskii, S. A.*: Chemistry and Physics of Agricultural Hydrogels. Vol. 104, pp. 97–134.
Kennedy, J. P. see Jacob, S.: Vol. 146, pp. 1–38.
Kennedy, J. P. see Majoros, I.: Vol. 112, pp. 1–113.

Kennedy, K. A., Roberts, G. W. and *DeSimone, J. M.*: Heterogeneous Polymerization of Fluoroolefins in Supercritical Carbon Dioxide. Vol. 175, pp. 329–346.
Khalatur, P. G. and *Khokhlov, A. R.*: Computer-Aided Conformation-Dependent Design of Copolymer Sequences. Vol. 195, pp. 1–100.
Khokhlov, A., Starodybtzev, S. and *Vasilevskaya, V.*: Conformational Transitions of Polymer Gels: Theory and Experiment. Vol. 109, pp. 121–172.
Khokhlov, A. R. see Grosberg, A. Y.: Vol. 196, pp. 189–210.
Khokhlov, A. R. see Khalatur, P. G.: Vol. 195, pp. 1–100.
Khokhlov, A. R. see Kuchanov, S. I.: Vol. 196, pp. 129–188.
Khokhlov, A. R. see Okhapkin, I. M.: Vol. 195, pp. 177–210.
Kiefer, J., Hedrick, J. L. and *Hiborn, J. G.*: Macroporous Thermosets by Chemically Induced Phase Separation. Vol. 147, pp. 161–247.
Kihara, N. see Takata, T.: Vol. 171, pp. 1–75.
Kilian, H. G. and *Pieper, T.*: Packing of Chain Segments. A Method for Describing X-Ray Patterns of Crystalline, Liquid Crystalline and Non-Crystalline Polymers. Vol. 108, pp. 49–90.
Kim, J. see Quirk, R. P.: Vol. 153, pp. 67–162.
Kim, K.-S. see Lin, T.-C.: Vol. 161, pp. 157–193.
Kimmich, R. and *Fatkullin, N.*: Polymer Chain Dynamics and NMR. Vol. 170, pp. 1–113.
Kippelen, B. and *Peyghambarian, N.*: Photorefractive Polymers and their Applications. Vol. 161, pp. 87–156.
Kirchhoff, R. A. and *Bruza, K. J.*: Polymers from Benzocyclobutenes. Vol. 117, pp. 1–66.
Kishore, K. and *Ganesh, K.*: Polymers Containing Disulfide, Tetrasulfide, Diselenide and Ditelluride Linkages in the Main Chain. Vol. 121, pp. 81–122.
Kitamaru, R.: Phase Structure of Polyethylene and Other Crystalline Polymers by Solid-State 13C/MNR. Vol. 137, pp. 41–102.
Klapper, M. see Rusanov, A. L.: Vol. 179, pp. 83–134.
Klee, D. and *Höcker, H.*: Polymers for Biomedical Applications: Improvement of the Interface Compatibility. Vol. 149, pp. 1–57.
Klemm, E., Pautzsch, T. and *Blankenburg, L.*: Organometallic PAEs. Vol. 177, pp. 53–90.
Klier, J. see Scranton, A. B.: Vol. 122, pp. 1–54.
v. Klitzing, R. and *Tieke, B.*: Polyelectrolyte Membranes. Vol. 165, pp. 177–210.
Kloeckner, J. see Wagner, E.: Vol. 192, pp. 135–173.
Klüppel, M.: The Role of Disorder in Filler Reinforcement of Elastomers on Various Length Scales. Vol. 164, pp. 1–86.
Klüppel, M. see Heinrich, G.: Vol. 160, pp. 1–44.
Knuuttila, H., Lehtinen, A. and *Nummila-Pakarinen, A.*: Advanced Polyethylene Technologies – Controlled Material Properties. Vol. 169, pp. 13–27.
Kobayashi, S. and *Ohmae, M.*: Enzymatic Polymerization to Polysaccharides. Vol. 194, pp. 159–210.
Kobayashi, S. see Uyama, H.: Vol. 194, pp. 51–67.
Kobayashi, S. see Uyama, H.: Vol. 194, pp. 133–158.
Kobayashi, S., Shoda, S. and *Uyama, H.*: Enzymatic Polymerization and Oligomerization. Vol. 121, pp. 1–30.
Kobayashi, T. see Abe, A.: Vol. 181, pp. 121–152.
Köhler, W. and *Schäfer, R.*: Polymer Analysis by Thermal-Diffusion Forced Rayleigh Scattering. Vol. 151, pp. 1–59.
Koenig, J. L. see Bhargava, R.: Vol. 163, pp. 137–191.
Koenig, J. L. see Andreis, M.: Vol. 124, pp. 191–238.
Koike, T.: Viscoelastic Behavior of Epoxy Resins Before Crosslinking. Vol. 148, pp. 139–188.

Kokko, E. see Löfgren, B.: Vol. 169, pp. 1–12.
Kokufuta, E.: Novel Applications for Stimulus-Sensitive Polymer Gels in the Preparation of Functional Immobilized Biocatalysts. Vol. 110, pp. 157–178.
Konishi, T. see Kaji, K.: Vol. 191, pp. 187–240.
Konno, M. see Saito, S.: Vol. 109, pp. 207–232.
Konradi, R. see Rühe, J.: Vol. 165, pp. 79–150.
Kopecek, J. see Putnam, D.: Vol. 122, pp. 55–124.
Koßmehl, G. see Schopf, G.: Vol. 129, pp. 1–145.
Kostoglodov, P. V. see Rusanov, A. L.: Vol. 179, pp. 83–134.
Kozlov, E. see Prokop, A.: Vol. 160, pp. 119–174.
Kramer, E. J. see Creton, C.: Vol. 156, pp. 53–135.
Kremer, K. see Baschnagel, J.: Vol. 152, pp. 41–156.
Kremer, K. see Holm, C.: Vol. 166, pp. 67–111.
Kressler, J. see Kammer, H. W.: Vol. 106, pp. 31–86.
Kricheldorf, H. R.: Liquid-Cristalline Polyimides. Vol. 141, pp. 83–188.
Krishnamoorti, R. see Giannelis, E. P.: Vol. 138, pp. 107–148.
Krüger, J. and *Kautek, W.*: Ultrashort Pulse Laser Interaction with Dielectrics and Polymers, Vol. 168, pp. 247–290.
Kuchanov, S. I.: Modern Aspects of Quantitative Theory of Free-Radical Copolymerization. Vol. 103, pp. 1–102.
Kuchanov, S. I. and *Khokhlov, A. R.*: Role of Physical Factors in the Process of Obtaining Copolymers. Vol. 196, pp. 129–188.
Kuchanov, S. I.: Principles of Quantitive Description of Chemical Structure of Synthetic Polymers. Vol. 152, pp. 157–202.
Kudaibergennow, S. E.: Recent Advances in Studying of Synthetic Polyampholytes in Solutions. Vol. 144, pp. 115–198.
Kuleznev, V. N. see Kandyrin, L. B.: Vol. 103, pp. 103–148.
Kulichkhin, S. G. see Malkin, A. Y.: Vol. 101, pp. 217–258.
Kulicke, W.-M. see Grigorescu, G.: Vol. 152, pp. 1–40.
Kumar, M. N. V. R., Kumar, N., Domb, A. J. and *Arora, M.*: Pharmaceutical Polymeric Controlled Drug Delivery Systems. Vol. 160, pp. 45–118.
Kumar, N. see Kumar, M. N. V. R.: Vol. 160, pp. 45–118.
Kummerloewe, C. see Kammer, H. W.: Vol. 106, pp. 31–86.
Kuznetsova, N. P. see Samsonov, G. V.: Vol. 104, pp. 1–50.
Kwon, Y. and *Faust, R.*: Synthesis of Polyisobutylene-Based Block Copolymers with Precisely Controlled Architecture by Living Cationic Polymerization. Vol. 167, pp. 107–135.

Labadie, J. W. see Hergenrother, P. M.: Vol. 117, pp. 67–110.
Labadie, J. W. see Hedrick, J. L.: Vol. 141, pp. 1–44.
Labadie, J. W. see Hedrick, J. L.: Vol. 147, pp. 61–112.
Lamparski, H. G. see O'Brien, D. F.: Vol. 126, pp. 53–84.
Laschewsky, A.: Molecular Concepts, Self-Organisation and Properties of Polysoaps. Vol. 124, pp. 1–86.
Laso, M. see Leontidis, E.: Vol. 116, pp. 283–318.
Lauprêtre, F. see Monnerie, L.: Vol. 187, pp. 35–213.
Lazár, M. and *Rychl, R.*: Oxidation of Hydrocarbon Polymers. Vol. 102, pp. 189–222.
Lechowicz, J. see Galina, H.: Vol. 137, pp. 135–172.
Léger, L., Raphaël, E. and *Hervet, H.*: Surface-Anchored Polymer Chains: Their Role in Adhesion and Friction. Vol. 138, pp. 185–226.
Lenz, R. W.: Biodegradable Polymers. Vol. 107, pp. 1–40.

Leontidis, E., de Pablo, J. J., Laso, M. and *Suter, U. W.*: A Critical Evaluation of Novel Algorithms for the Off-Lattice Monte Carlo Simulation of Condensed Polymer Phases. Vol. 116, pp. 283–318.
Lee, B. see Quirk, R. P.: Vol. 153, pp. 67–162.
Lee, K.-S. see Kajzar, F.: Vol. 161, pp. 1–85.
Lee, Y. see Quirk, R. P.: Vol. 153, pp. 67–162.
Lehtinen, A. see Knuuttila, H.: Vol. 169, pp. 13–27.
Leónard, D. see Mathieu, H. J.: Vol. 162, pp. 1–35.
Lesec, J. see Viovy, J.-L.: Vol. 114, pp. 1–42.
Levesque, D. see Weis, J.-J.: Vol. 185, pp. 163–225.
Li, L. and *de Jeu, W. H.*: Flow-induced mesophases in crystallizable polymers. Vol. 181, pp. 75–120.
Li, L. see Chan, C.-M.: Vol. 188, pp. 1–41.
Li, M., Coenjarts, C. and *Ober, C. K.*: Patternable Block Copolymers. Vol. 190, pp. 183–226.
Li, M. see Jiang, M.: Vol. 146, pp. 121–194.
Liang, G. L. see Sumpter, B. G.: Vol. 116, pp. 27–72.
Lienert, K.-W.: Poly(ester-imide)s for Industrial Use. Vol. 141, pp. 45–82.
Likhatchev, D. see Rusanov, A. L.: Vol. 179, pp. 83–134.
Lin, J. and *Sherrington, D. C.*: Recent Developments in the Synthesis, Thermostability and Liquid Crystal Properties of Aromatic Polyamides. Vol. 111, pp. 177–220.
Lin, T.-C., Chung, S.-J., Kim, K.-S., Wang, X., He, G. S., Swiatkiewicz, J., Pudavar, H. E. and *Prasad, P. N.*: Organics and Polymers with High Two-Photon Activities and their Applications. Vol. 161, pp. 157–193.
Linse, P.: Simulation of Charged Colloids in Solution. Vol. 185, pp. 111–162.
Lippert, T.: Laser Application of Polymers. Vol. 168, pp. 51–246.
Liu, Y. see Söderqvist Lindblad, M.: Vol. 157, pp. 139–161.
Long, T.-C. see Geil, P. H.: Vol. 180, pp. 89–159.
López Cabarcos, E. see Baltá-Calleja, F. J.: Vol. 108, pp. 1–48.
Lotz, B.: Analysis and Observation of Polymer Crystal Structures at the Individual Stem Level. Vol. 180, pp. 17–44.
Löfgren, B., Kokko, E. and *Seppälä, J.*: Specific Structures Enabled by Metallocene Catalysis in Polyethenes. Vol. 169, pp. 1–12.
Löwen, H. see Thünemann, A. F.: Vol. 166, pp. 113–171.
Lozinsky V. I.: Approaches to Chemical Synthesis of Protein-Like Copolymers. Vol. 196, pp. 87–128.
Luo, Y. see Schork, F. J.: Vol. 175, pp. 129–255.

Macko, T. and *Hunkeler, D.*: Liquid Chromatography under Critical and Limiting Conditions: A Survey of Experimental Systems for Synthetic Polymers. Vol. 163, pp. 61–136.
Maeda, H., Greish, K. and *Fang, J.*: The EPR Effect and Polymeric Drugs: A Paradigm Shift for Cancer Chemotherapy in the 21st Century. Vol. 193, pp. 103–121.
Majoros, I., Nagy, A. and *Kennedy, J. P.*: Conventional and Living Carbocationic Polymerizations United. I. A Comprehensive Model and New Diagnostic Method to Probe the Mechanism of Homopolymerizations. Vol. 112, pp. 1–113.
Makhaeva, E. E. see Okhapkin, I. M.: Vol. 195, pp. 177–210.
Makhija, S. see Jaffe, M.: Vol. 117, pp. 297–328.
Malmström, E. see Hult, A.: Vol. 143, pp. 1–34.
Malkin, A. Y. and *Kulichkhin, S. G.*: Rheokinetics of Curing. Vol. 101, pp. 217–258.
Maniar, M. see Domb, A. J.: Vol. 107, pp. 93–142.
Manias, E. see Giannelis, E. P.: Vol. 138, pp. 107–148.

Martin, H. see Engelhardt, H.: Vol. 165, pp. 211–247.
Marty, J. D. and *Mauzac, M.*: Molecular Imprinting: State of the Art and Perspectives. Vol. 172, pp. 1–35.
Mashima, K., Nakayama, Y. and *Nakamura, A.*: Recent Trends in Polymerization of a-Olefins Catalyzed by Organometallic Complexes of Early Transition Metals. Vol. 133, pp. 1–52.
Mathew, D. see Reghunadhan Nair, C. P.: Vol. 155, pp. 1–99.
Mathieu, H. J., Chevolot, Y, Ruiz-Taylor, L. and *Leónard, D.*: Engineering and Characterization of Polymer Surfaces for Biomedical Applications. Vol. 162, pp. 1–35.
Matsuba, G. see Kaji, K.: Vol. 191, pp. 187–240.
Matsuda T.: Photoiniferter-Driven Precision Surface Graft Microarchitectures for Biomedical Applications. Vol. 197, pp. 67–106.
Matsumura S.: Enzymatic Synthesis of Polyesters via Ring-Opening Polymerization. Vol. 194, pp. 95–132.
Matsumoto, A.: Free-Radical Crosslinking Polymerization and Copolymerization of Multivinyl Compounds. Vol. 123, pp. 41–80.
Matsumoto, A. see Otsu, T.: Vol. 136, pp. 75–138.
Matsuoka, H. and *Ise, N.*: Small-Angle and Ultra-Small Angle Scattering Study of the Ordered Structure in Polyelectrolyte Solutions and Colloidal Dispersions. Vol. 114, pp. 187–232.
Matsushige, K., Hiramatsu, N. and *Okabe, H.*: Ultrasonic Spectroscopy for Polymeric Materials. Vol. 125, pp. 147–186.
Mattice, W. L. see Rehahn, M.: Vol. 131/132, pp. 1–475.
Mattice, W. L. see Baschnagel, J.: Vol. 152, pp. 41–156.
Mattozzi, A. see Gedde, U. W.: Vol. 169, pp. 29–73.
Mauzac, M. see Marty, J. D.: Vol. 172, pp. 1–35.
Mays, W. see Xu, Z.: Vol. 120, pp. 1–50.
Mays, J. W. see Pitsikalis, M.: Vol. 135, pp. 1–138.
McGrath, J. E. see Hedrick, J. L.: Vol. 141, pp. 1–44.
McGrath, J. E., Dunson, D. L. and *Hedrick, J. L.*: Synthesis and Characterization of Segmented Polyimide-Polyorganosiloxane Copolymers. Vol. 140, pp. 61–106.
McLeish, T. C. B. and *Milner, S. T.*: Entangled Dynamics and Melt Flow of Branched Polymers. Vol. 143, pp. 195–256.
Mecerreyes, D., Dubois, P. and *Jerome, R.*: Novel Macromolecular Architectures Based on Aliphatic Polyesters: Relevance of the Coordination-Insertion Ring-Opening Polymerization. Vol. 147, pp. 1–60.
Mecham, S. J. see McGrath, J. E.: Vol. 140, pp. 61–106.
Meille, S. V. see Allegra, G.: Vol. 191, pp. 87–135.
Menzel, H. see Möhwald, H.: Vol. 165, pp. 151–175.
Meyer, T. see Spange, S.: Vol. 165, pp. 43–78.
Michler, G. H. see Kausch, H.-H.: Vol. 187, pp. 1–33.
Mikos, A. G. see Thomson, R. C.: Vol. 122, pp. 245–274.
Milner, S. T. see McLeish, T. C. B.: Vol. 143, pp. 195–256.
Mison, P. and *Sillion, B.*: Thermosetting Oligomers Containing Maleimides and Nadiimides End-Groups. Vol. 140, pp. 137–180.
Miyasaka, K.: PVA-Iodine Complexes: Formation, Structure and Properties. Vol. 108, pp. 91–130.
Miller, R. D. see Hedrick, J. L.: Vol. 141, pp. 1–44.
Minko, S. see Rühe, J.: Vol. 165, pp. 79–150.
Möhwald, H., Menzel, H., Helm, C. A. and *Stamm, M.*: Lipid and Polyampholyte Monolayers to Study Polyelectrolyte Interactions and Structure at Interfaces. Vol. 165, pp. 151–175.

Monkenbusch, M. see *Richter, D.*: Vol. 174, pp. 1–221.
Monnerie, L., Halary, J. L. and *Kausch, H.-H.*: Deformation, Yield and Fracture of Amorphous Polymers: Relation to the Secondary Transitions. Vol. 187, pp. 215–364.
Monnerie, L., Lauprêtre, F. and *Halary, J. L.*: Investigation of Solid-State Transitions in Linear and Crosslinked Amorphous Polymers. Vol. 187, pp. 35–213.
Monnerie, L. see *Bahar, I.*: Vol. 116, pp. 145–206.
Moore, J. S. see *Ray, C. R.*: Vol. 177, pp. 99–149.
Mori, H. see *Bohrisch, J.*: Vol. 165, pp. 1–41.
Morishima, Y.: Photoinduced Electron Transfer in Amphiphilic Polyelectrolyte Systems. Vol. 104, pp. 51–96.
Morton, M. see *Quirk, R. P.*: Vol. 153, pp. 67–162.
Motornov, M. see *Rühe, J.*: Vol. 165, pp. 79–150.
Mours, M. see *Winter, H. H.*: Vol. 134, pp. 165–234.
Müllen, K. see *Scherf, U.*: Vol. 123, pp. 1–40.
Müller, A. H. E. see *Bohrisch, J.*: Vol. 165, pp. 1–41.
Müller, A. H. E. see *Förster, S.*: Vol. 166, pp. 173–210.
Müller, A. J., Balsamo, V. and *Arnal, M. L.*: Nucleation and Crystallization in Diblock and Triblock Copolymers. Vol. 190, pp. 1–63.
Müller, M. and *Schmid, F.*: Incorporating Fluctuations and Dynamics in Self-Consistent Field Theories for Polymer Blends. Vol. 185, pp. 1–58.
Müller, M. see *Thünemann, A. F.*: Vol. 166, pp. 113–171.
Müller-Plathe, F. see *Gusev, A. A.*: Vol. 116, pp. 207–248.
Müller-Plathe, F. see *Baschnagel, J.*: Vol. 152, p. 41–156.
Mukerherjee, A. see *Biswas, M.*: Vol. 115, pp. 89–124.
Munz, M., Cappella, B., Sturm, H., Geuss, M. and *Schulz, E.*: Materials Contrasts and Nanolithography Techniques in Scanning Force Microscopy (SFM) and their Application to Polymers and Polymer Composites. Vol. 164, pp. 87–210.
Murat, M. see *Baschnagel, J.*: Vol. 152, p. 41–156.
Muthukumar, M.: Modeling Polymer Crystallization. Vol. 191, pp. 241–274.
Muzzarelli, C. see *Muzzarelli, R. A. A.*: Vol. 186, pp. 151–209.
Muzzarelli, R. A. A. and *Muzzarelli, C.*: Chitosan Chemistry: Relevance to the Biomedical Sciences. Vol. 186, pp. 151–209.
Mylnikov, V.: Photoconducting Polymers. Vol. 115, pp. 1–88.

Nagy, A. see *Majoros, I.*: Vol. 112, pp. 1–11.
Naka, K. see *Uemura, T.*: Vol. 167, pp. 81–106.
Nakamura, A. see *Mashima, K.*: Vol. 133, pp. 1–52.
Nakayama, Y. see *Mashima, K.*: Vol. 133, pp. 1–52.
Narasinham, B. and *Peppas, N. A.*: The Physics of Polymer Dissolution: Modeling Approaches and Experimental Behavior. Vol. 128, pp. 157–208.
Nechaev, S. see *Grosberg, A.*: Vol. 106, pp. 1–30.
Neoh, K. G. see *Kang, E. T.*: Vol. 106, pp. 135–190.
Netz, R. R. see *Holm, C.*: Vol. 166, pp. 67–111.
Netz, R. R. see *Rühe, J.*: Vol. 165, pp. 79–150.
Newman, S. M. see *Anseth, K. S.*: Vol. 122, pp. 177–218.
Nijenhuis, K. te: Thermoreversible Networks. Vol. 130, pp. 1–252.
Ninan, K. N. see *Reghunadhan Nair, C. P.*: Vol. 155, pp. 1–99.
Nishi, T. see *Jinnai, H.*: Vol. 170, pp. 115–167.
Nishida, K. see *Kaji, K.*: Vol. 191, pp. 187–240.
Nishikawa, Y. see *Jinnai, H.*: Vol. 170, pp. 115–167.

Nishiyama, N. and *Kataoka, K.*: Nanostructured Devices Based on Block Copolymer Assemblies for Drug Delivery: Designing Structures for Enhanced Drug Function. Vol. 193, pp. 67–101.
Noid, D. W. see Otaigbe, J. U.: Vol. 154, pp. 1–86.
Noid, D. W. see Sumpter, B. G.: Vol. 116, pp. 27–72.
Nomura, M., Tobita, H. and *Suzuki, K.*: Emulsion Polymerization: Kinetic and Mechanistic Aspects. Vol. 175, pp. 1–128.
Northolt, M. G., Picken, S. J., Den Decker, M. G., Baltussen, J. J. M. and *Schlatmann, R.*: The Tensile Strength of Polymer Fibres. Vol. 178, pp. 1–108.
Novac, B. see Grubbs, R.: Vol. 102, pp. 47–72.
Novikov, V. V. see Privalko, V. P.: Vol. 119, pp. 31–78.
Nummila-Pakarinen, A. see Knuuttila, H.: Vol. 169, pp. 13–27.

Ober, C. K. see Li, M.: Vol. 190, pp. 183–226.
O'Brien, D. F., Armitage, B. A., Bennett, D. E. and *Lamparski, H. G.*: Polymerization and Domain Formation in Lipid Assemblies. Vol. 126, pp. 53–84.
Ogasawara, M.: Application of Pulse Radiolysis to the Study of Polymers and Polymerizations. Vol. 105, pp. 37–80.
Ohmae, M. see Kobayashi, S.: Vol. 194, pp. 159–210.
Ohno, K. see Tsujii, Y.: Vol. 197, pp. 1–47.
Okabe, H. see Matsushige, K.: Vol. 125, pp. 147–186.
Okada, M.: Ring-Opening Polymerization of Bicyclic and Spiro Compounds. Reactivities and Polymerization Mechanisms. Vol. 102, pp. 1–46.
Okada, K. see Hikosaka, M.: Vol. 191, pp. 137–186.
Okano, T.: Molecular Design of Temperature-Responsive Polymers as Intelligent Materials. Vol. 110, pp. 179–198.
Okay, O. see Funke, W.: Vol. 136, pp. 137–232.
Okhapkin, I. M., Makhaeva, E. E. and *Khokhlov, A. R.*: Water Solutions of Amphiphilic Polymers: Nanostructure Formation and Possibilities for Catalysis. Vol. 195, pp. 177–210.
Onuki, A.: Theory of Phase Transition in Polymer Gels. Vol. 109, pp. 63–120.
Oppermann, W. see Holm, C.: Vol. 166, pp. 1–27.
Oppermann, W. see Volk, N.: Vol. 166, pp. 29–65.
Osad'ko, I. S.: Selective Spectroscopy of Chromophore Doped Polymers and Glasses. Vol. 114, pp. 123–186.
Osakada, K. and *Takeuchi, D.*: Coordination Polymerization of Dienes, Allenes, and Methylenecycloalkanes. Vol. 171, pp. 137–194.
Otaigbe, J. U., Barnes, M. D., Fukui, K., Sumpter, B. G. and *Noid, D. W.*: Generation, Characterization, and Modeling of Polymer Micro- and Nano-Particles. Vol. 154, pp. 1–86.
Otsu, T. and *Matsumoto, A.*: Controlled Synthesis of Polymers Using the Iniferter Technique: Developments in Living Radical Polymerization. Vol. 136, pp. 75–138.

de Pablo, J. J. see Leontidis, E.: Vol. 116, pp. 283–318.
Padias, A. B. see Penelle, J.: Vol. 102, pp. 73–104.
Pascault, J.-P. see Williams, R. J. J.: Vol. 128, pp. 95–156.
Pasch, H.: Analysis of Complex Polymers by Interaction Chromatography. Vol. 128, pp. 1–46.
Pasch, H.: Hyphenated Techniques in Liquid Chromatography of Polymers. Vol. 150, pp. 1–66.
Pasut, G. and *Veronese, F. M.*: PEGylation of Proteins as Tailored Chemistry for Optimized Bioconjugates. Vol. 192, pp. 95–134.
Paul, W. see Baschnagel, J.: Vol. 152, pp. 41–156.

Paulsen, S. B. and *Barsett, H.*: Bioactive Pectic Polysaccharides. Vol. 186, pp. 69–101.
Pautzsch, T. see Klemm, E.: Vol. 177, pp. 53–90.
Penczek, P., Czub, P. and *Pielichowski, J.*: Unsaturated Polyester Resins: Chemistry and Technology. Vol. 184, pp. 1–95.
Penczek, P. see Batog, A. E.: Vol. 144, pp. 49–114.
Penczek, P. see Bogdal, D.: Vol. 163, pp. 193–263.
Penelle, J., Hall, H. K., Padias, A. B. and *Tanaka, H.*: Captodative Olefins in Polymer Chemistry. Vol. 102, pp. 73–104.
Peppas, N. A. see Bell, C. L.: Vol. 122, pp. 125–176.
Peppas, N. A. see Hassan, C. M.: Vol. 153, pp. 37–65.
Peppas, N. A. see Narasimhan, B.: Vol. 128, pp. 157–208.
Petersen, K. L. see Geil, P. H.: Vol. 180, pp. 89–159.
Pet'ko, I. P. see Batog, A. E.: Vol. 144, pp. 49–114.
Pheyghambarian, N. see Kippelen, B.: Vol. 161, pp. 87–156.
Pichot, C. see Hunkeler, D.: Vol. 112, pp. 115–134.
Picken, S. J. see Northolt, M. G.: Vol. 178, pp. 1–108.
Pielichowski, J. see Bogdal, D.: Vol. 163, pp. 193–263.
Pielichowski, J. see Penczek, P.: Vol. 184, pp. 1–95.
Pieper, T. see Kilian, H. G.: Vol. 108, pp. 49–90.
Pispas, S. see Pitsikalis, M.: Vol. 135, pp. 1–138.
Pispas, S. see Hadjichristidis, N.: Vol. 142, pp. 71–128.
Pitsikalis, M., Pispas, S., Mays, J. W. and *Hadjichristidis, N.*: Nonlinear Block Copolymer Architectures. Vol. 135, pp. 1–138.
Pitsikalis, M. see Hadjichristidis, N.: Vol. 142, pp. 71–128.
Pitsikalis, M. see Hadjichristidis, N.: Vol. 189, pp. 1–124.
Pleul, D. see Spange, S.: Vol. 165, pp. 43–78.
Plummer, C. J. G.: Microdeformation and Fracture in Bulk Polyolefins. Vol. 169, pp. 75–119.
Pötschke, D. see Dingenouts, N.: Vol. 144, pp. 1–48.
Pokrovskii, V. N.: The Mesoscopic Theory of the Slow Relaxation of Linear Macromolecules. Vol. 154, pp. 143–219.
Pospíšil, J.: Functionalized Oligomers and Polymers as Stabilizers for Conventional Polymers. Vol. 101, pp. 65–168.
Pospíšil, J.: Aromatic and Heterocyclic Amines in Polymer Stabilization. Vol. 124, pp. 87–190.
Powers, A. C. see Prokop, A.: Vol. 136, pp. 53–74.
Prasad, P. N. see Lin, T.-C.: Vol. 161, pp. 157–193.
Priddy, D. B.: Recent Advances in Styrene Polymerization. Vol. 111, pp. 67–114.
Priddy, D. B.: Thermal Discoloration Chemistry of Styrene-co-Acrylonitrile. Vol. 121, pp. 123–154.
Privalko, V. P. and *Novikov, V. V.*: Model Treatments of the Heat Conductivity of Heterogeneous Polymers. Vol. 119, pp. 31–78.
Prociak, A. see Bogdal, D.: Vol. 163, pp. 193–263.
Prokop, A., Hunkeler, D., DiMari, S., Haralson, M. A. and *Wang, T. G.*: Water Soluble Polymers for Immunoisolation I: Complex Coacervation and Cytotoxicity. Vol. 136, pp. 1–52.
Prokop, A., Hunkeler, D., Powers, A. C., Whitesell, R. R. and *Wang, T. G.*: Water Soluble Polymers for Immunoisolation II: Evaluation of Multicomponent Microencapsulation Systems. Vol. 136, pp. 53–74.
Prokop, A., Kozlov, E., Carlesso, G. and *Davidsen, J. M.*: Hydrogel-Based Colloidal Polymeric System for Protein and Drug Delivery: Physical and Chemical Characterization, Permeability Control and Applications. Vol. 160, pp. 119–174.

Pruitt, L. A.: The Effects of Radiation on the Structural and Mechanical Properties of Medical Polymers. Vol. 162, pp. 65–95.
Pudavar, H. E. see Lin, T.-C.: Vol. 161, pp. 157–193.
Pukánszky, B. and *Fekete, E.*: Adhesion and Surface Modification. Vol. 139, pp. 109–154.
Putnam, D. and *Kopecek, J.*: Polymer Conjugates with Anticancer Acitivity. Vol. 122, pp. 55–124.
Putra, E. G. R. see Ungar, G.: Vol. 180, pp. 45–87.

Quirk, R. P., Yoo, T., Lee, Y., M., Kim, J. and *Lee, B.*: Applications of 1,1-Diphenylethylene Chemistry in Anionic Synthesis of Polymers with Controlled Structures. Vol. 153, pp. 67–162.

Ramaraj, R. and *Kaneko, M.*: Metal Complex in Polymer Membrane as a Model for Photosynthetic Oxygen Evolving Center. Vol. 123, pp. 215–242.
Rangarajan, B. see Scranton, A. B.: Vol. 122, pp. 1–54.
Ranucci, E. see Söderqvist Lindblad, M.: Vol. 157, pp. 139–161.
Raphaël, E. see Léger, L.: Vol. 138, pp. 185–226.
Rastogi, S. and *Terry, A. E.*: Morphological implications of the interphase bridging crystalline and amorphous regions in semi-crystalline polymers. Vol. 180, pp. 161–194.
Ray, C. R. and *Moore, J. S.*: Supramolecular Organization of Foldable Phenylene Ethynylene Oligomers. Vol. 177, pp. 99–149.
Reddinger, J. L. and *Reynolds, J. R.*: Molecular Engineering of p-Conjugated Polymers. Vol. 145, pp. 57–122.
Reghunadhan Nair, C. P., Mathew, D. and *Ninan, K. N.*: Cyanate Ester Resins, Recent Developments. Vol. 155, pp. 1–99.
Reichert, K. H. see Hunkeler, D.: Vol. 112, pp. 115–134.
Reihmann, M. and *Ritter, H.*: Synthesis of Phenol Polymers Using Peroxidases. Vol. 194, pp. 1–49.
Rehahn, M., Mattice, W. L. and *Suter, U. W.*: Rotational Isomeric State Models in Macromolecular Systems. Vol. 131/132, pp. 1–475.
Rehahn, M. see Bohrisch, J.: Vol. 165, pp. 1–41.
Rehahn, M. see Holm, C.: Vol. 166, pp. 1–27.
Reineker, P. see Holm, C.: Vol. 166, pp. 67–111.
Reitberger, T. see Jacobson, K.: Vol. 169, pp. 151–176.
Ritter, H. see Reihmann, M.: Vol. 194, pp. 1–49.
Reynolds, J. R. see Reddinger, J. L.: Vol. 145, pp. 57–122.
Richter, D. see Ewen, B.: Vol. 134, pp. 1–130.
Richter, D., Monkenbusch, M. and *Colmenero, J.*: Neutron Spin Echo in Polymer Systems. Vol. 174, pp. 1–221.
Riegler, S. see Trimmel, G.: Vol. 176, pp. 43–87.
Ringsdorf, H. see Duncan, R.: Vol. 192, pp. 1–8.
Risse, W. see Grubbs, R.: Vol. 102, pp. 47–72.
Rivas, B. L. and *Geckeler, K. E.*: Synthesis and Metal Complexation of Poly(ethyleneimine) and Derivatives. Vol. 102, pp. 171–188.
Roberts, C. J. see Ellis, J. S.: Vol. 193, pp. 123–172.
Roberts, G. W. see Kennedy, K. A.: Vol. 175, pp. 329–346.
Robin, J. J.: The Use of Ozone in the Synthesis of New Polymers and the Modification of Polymers. Vol. 167, pp. 35–79.
Robin, J. J. see Boutevin, B.: Vol. 102, pp. 105–132.

Rodríguez-Pérez, M. A.: Crosslinked Polyolefin Foams: Production, Structure, Properties, and Applications. Vol. 184, pp. 97–126.
Roe, R.-J.: MD Simulation Study of Glass Transition and Short Time Dynamics in Polymer Liquids. Vol. 116, pp. 111–114.
Roovers, J. and *Comanita, B.*: Dendrimers and Dendrimer-Polymer Hybrids. Vol. 142, pp. 179–228.
Rothon, R. N.: Mineral Fillers in Thermoplastics: Filler Manufacture and Characterisation. Vol. 139, pp. 67–108.
de Rosa, C. see Auriemma, F.: Vol. 181, pp. 1–74.
Rozenberg, B. A. see Williams, R. J. J.: Vol. 128, pp. 95–156.
Rühe, J., Ballauff, M., Biesalski, M., Dziezok, P., Gröhn, F., Johannsmann, D., Houbenov, N., Hugenberg, N., Konradi, R., Minko, S., Motornov, M., Netz, R. R., Schmidt, M., Seidel, C., Stamm, M., Stephan, T., Usov, D. and *Zhang, H.*: Polyelectrolyte Brushes. Vol. 165, pp. 79–150.
Ruckenstein, E.: Concentrated Emulsion Polymerization. Vol. 127, pp. 1–58.
Ruiz-Taylor, L. see Mathieu, H. J.: Vol. 162, pp. 1–35.
Rusanov, A. L.: Novel Bis (Naphtalic Anhydrides) and Their Polyheteroarylenes with Improved Processability. Vol. 111, pp. 115–176.
Rusanov, A. L., Likhatchev, D., Kostoglodov, P. V., Müllen, K. and *Klapper, M.*: Proton-Exchanging Electrolyte Membranes Based on Aromatic Condensation Polymers. Vol. 179, pp. 83–134.
Russel, T. P. see Hedrick, J. L.: Vol. 141, pp. 1–44.
Russum, J. P. see Schork, F. J.: Vol. 175, pp. 129–255.
Rychly, J. see Lazár, M.: Vol. 102, pp. 189–222.
Ryner, M. see Stridsberg, K. M.: Vol. 157, pp. 27–51.
Ryzhov, V. A. see Bershtein, V. A.: Vol. 114, pp. 43–122.

Sabsai, O. Y. see Barshtein, G. R.: Vol. 101, pp. 1–28.
Saburov, V. V. see Zubov, V. P.: Vol. 104, pp. 135–176.
Saito, S., Konno, M. and *Inomata, H.*: Volume Phase Transition of N-Alkylacrylamide Gels. Vol. 109, pp. 207–232.
Samsonov, G. V. and *Kuznetsova, N. P.*: Crosslinked Polyelectrolytes in Biology. Vol. 104, pp. 1–50.
Santa Cruz, C. see Baltá-Calleja, F. J.: Vol. 108, pp. 1–48.
Santos, S. see Baschnagel, J.: Vol. 152, p. 41–156.
Satchi-Fainaro, R., Duncan, R. and *Barnes, C. M.*: Polymer Therapeutics for Cancer: Current Status and Future Challenges. Vol. 193, pp. 1–65.
Satchi-Fainaro, R. see Duncan, R.: Vol. 192, pp. 1–8.
Sato, T. and *Teramoto, A.*: Concentrated Solutions of Liquid-Christalline Polymers. Vol. 126, pp. 85–162.
Schaller, C. see Bohrisch, J.: Vol. 165, pp. 1–41.
Schäfer, R. see Köhler, W.: Vol. 151, pp. 1–59.
Scherf, U. and *Müllen, K.*: The Synthesis of Ladder Polymers. Vol. 123, pp. 1–40.
Sherman, S. see Kabanov, A. V.: Vol. 193, pp. 173–198.
Schlatmann, R. see Northolt, M. G.: Vol. 178, pp. 1–108.
Schmid, F. see Müller, M.: Vol. 185, pp. 1–58.
Schmidt, M. see Förster, S.: Vol. 120, pp. 51–134.
Schmidt, M. see Rühe, J.: Vol. 165, pp. 79–150.
Schmidt, M. see Volk, N.: Vol. 166, pp. 29–65.
Scholz, M.: Effects of Ion Radiation on Cells and Tissues. Vol. 162, pp. 97–158.

Schönherr, H. see Vancso, G. J.: Vol. 182, pp. 55–129.
Schopf, G. and *Koßmehl, G.*: Polythiophenes – Electrically Conductive Polymers. Vol. 129, pp. 1–145.
Schork, F. J., Luo, Y., Smulders, W., Russum, J. P., Butté, A. and *Fontenot, K.*: Miniemulsion Polymerization. Vol. 175, pp. 127–255.
Schulz, E. see Munz, M.: Vol. 164, pp. 97–210.
Schwahn, D.: Critical to Mean Field Crossover in Polymer Blends. Vol. 183, pp. 1–61.
Seppälä, J. see Löfgren, B.: Vol. 169, pp. 1–12.
Sturm, H. see Munz, M.: Vol. 164, pp. 87–210.
Schweizer, K. S.: Prism Theory of the Structure, Thermodynamics, and Phase Transitions of Polymer Liquids and Alloys. Vol. 116, pp. 319–378.
Scranton, A. B., Rangarajan, B. and *Klier, J.*: Biomedical Applications of Polyelectrolytes. Vol. 122, pp. 1–54.
Sefton, M. V. and *Stevenson, W. T. K.*: Microencapsulation of Live Animal Cells Using Polycrylates. Vol. 107, pp. 143–198.
Seidel, C. see Holm, C.: Vol. 166, pp. 67–111.
Seidel, C. see Rühe, J.: Vol. 165, pp. 79–150.
El Seoud, O. A. and *Heinze, T.*: Organic Esters of Cellulose: New Perspectives for Old Polymers. Vol. 186, pp. 103–149.
Shabat, D. see Amir, R. J.: Vol. 192, pp. 59–94.
Shamanin, V. V.: Bases of the Axiomatic Theory of Addition Polymerization. Vol. 112, pp. 135–180.
Shcherbina, M. A. see Ungar, G.: Vol. 180, pp. 45–87.
Sheiko, S. S.: Imaging of Polymers Using Scanning Force Microscopy: From Superstructures to Individual Molecules. Vol. 151, pp. 61–174.
Sherrington, D. C. see Cameron, N. R.: Vol. 126, pp. 163–214.
Sherrington, D. C. see Lin, J.: Vol. 111, pp. 177–220.
Sherrington, D. C. see Steinke, J.: Vol. 123, pp. 81–126.
Shibayama, M. see Tanaka, T.: Vol. 109, pp. 1–62.
Shiga, T.: Deformation and Viscoelastic Behavior of Polymer Gels in Electric Fields. Vol. 134, pp. 131–164.
Shim, H.-K. and *Jin, J.*: Light-Emitting Characteristics of Conjugated Polymers. Vol. 158, pp. 191–241.
Shoda, S. see Kobayashi, S.: Vol. 121, pp. 1–30.
Siegel, R. A.: Hydrophobic Weak Polyelectrolyte Gels: Studies of Swelling Equilibria and Kinetics. Vol. 109, pp. 233–268.
de Silva, D. S. M. see Ungar, G.: Vol. 180, pp. 45–87.
Silvestre, F. see Calmon-Decriaud, A.: Vol. 207, pp. 207–226.
Sillion, B. see Mison, P.: Vol. 140, pp. 137–180.
Simon, F. see Spange, S.: Vol. 165, pp. 43–78.
Simon, G. P. see Becker, O.: Vol. 179, pp. 29–82.
Simon, P. F. W. see Abetz, V.: Vol. 189, pp. 125–212.
Simonutti, R. see Sozzani, P.: Vol. 181, pp. 153–177.
Singh, A. and *Kaplan, D. L.*: In Vitro Enzyme-Induced Vinyl Polymerization. Vol. 194, pp. 211–224.
Singh, A. see Xu, P.: Vol. 194, pp. 69–94.
Singh, R. P. see Sivaram, S.: Vol. 101, pp. 169–216.
Singh, R. P. see Desai, S. M.: Vol. 169, pp. 231–293.
Sinha Ray, S. see Biswas, M.: Vol. 155, pp. 167–221.

Sivaram, S. and *Singh, R. P.*: Degradation and Stabilization of Ethylene-Propylene Copolymers and Their Blends: A Critical Review. Vol. 101, pp. 169–216.
Slugovc, C. see Trimmel, G.: Vol. 176, pp. 43–87.
Smulders, W. see Schork, F. J.: Vol. 175, pp. 129–255.
Soares, J. B. P. see Anantawaraskul, S.: Vol. 182, pp. 1–54.
Sozzani, P., Bracco, S., Comotti, A. and *Simonutti, R.*: Motional Phase Disorder of Polymer Chains as Crystallized to Hexagonal Lattices. Vol. 181, pp. 153–177.
Söderqvist Lindblad, M., Liu, Y., Albertsson, A.-C., Ranucci, E. and *Karlsson, S.*: Polymer from Renewable Resources. Vol. 157, pp. 139–161.
Spange, S., Meyer, T., Voigt, I., Eschner, M., Estel, K., Pleul, D. and *Simon, F.*: Poly(Vinylformamide-co-Vinylamine)/Inorganic Oxid Hybrid Materials. Vol. 165, pp. 43–78.
Stamm, M. see Möhwald, H.: Vol. 165, pp. 151–175.
Stamm, M. see Rühe, J.: Vol. 165, pp. 79–150.
Starodybtzev, S. see Khokhlov, A.: Vol. 109, pp. 121–172.
Stegeman, G. I. see Canva, M.: Vol. 158, pp. 87–121.
Steinke, J., Sherrington, D. C. and *Dunkin, I. R.*: Imprinting of Synthetic Polymers Using Molecular Templates. Vol. 123, pp. 81–126.
Stelzer, F. see Trimmel, G.: Vol. 176, pp. 43–87.
Stenberg, B. see Jacobson, K.: Vol. 169, pp. 151–176.
Stenzenberger, H. D.: Addition Polyimides. Vol. 117, pp. 165–220.
Stephan, T. see Rühe, J.: Vol. 165, pp. 79–150.
Stevenson, W. T. K. see Sefton, M. V.: Vol. 107, pp. 143–198.
Stridsberg, K. M., Ryner, M. and *Albertsson, A.-C.*: Controlled Ring-Opening Polymerization: Polymers with Designed Macromoleculars Architecture. Vol. 157, pp. 27–51.
Sturm, H. see Munz, M.: Vol. 164, pp. 87–210.
Suematsu, K.: Recent Progress of Gel Theory: Ring, Excluded Volume, and Dimension. Vol. 156, pp. 136–214.
Sugimoto, H. and *Inoue, S.*: Polymerization by Metalloporphyrin and Related Complexes. Vol. 146, pp. 39–120.
Suginome, M. and *Ito, Y.*: Transition Metal-Mediated Polymerization of Isocyanides. Vol. 171, pp. 77–136.
Sumpter, B. G., Noid, D. W., Liang, G. L. and *Wunderlich, B.*: Atomistic Dynamics of Macromolecular Crystals. Vol. 116, pp. 27–72.
Sumpter, B. G. see Otaigbe, J. U.: Vol. 154, pp. 1–86.
Sun, H.-B. and *Kawata, S.*: Two-Photon Photopolymerization and 3D Lithographic Microfabrication. Vol. 170, pp. 169–273.
Suter, U. W. see Gusev, A. A.: Vol. 116, pp. 207–248.
Suter, U. W. see Leontidis, E.: Vol. 116, pp. 283–318.
Suter, U. W. see Rehahn, M.: Vol. 131/132, pp. 1–475.
Suter, U. W. see Baschnagel, J.: Vol. 152, pp. 41–156.
Suzuki, A.: Phase Transition in Gels of Sub-Millimeter Size Induced by Interaction with Stimuli. Vol. 110, pp. 199–240.
Suzuki, A. and *Hirasa, O.*: An Approach to Artifical Muscle by Polymer Gels due to Micro-Phase Separation. Vol. 110, pp. 241–262.
Suzuki, K. see Nomura, M.: Vol. 175, pp. 1–128.
Swiatkiewicz, J. see Lin, T.-C.: Vol. 161, pp. 157–193.

Tagawa, S.: Radiation Effects on Ion Beams on Polymers. Vol. 105, pp. 99–116.
Taguet, A., Ameduri, B. and *Boutevin, B.*: Crosslinking of Vinylidene Fluoride-Containing Fluoropolymers. Vol. 184, pp. 127–211.

Takata, T., Kihara, N. and *Furusho, Y.*: Polyrotaxanes and Polycatenanes: Recent Advances in Syntheses and Applications of Polymers Comprising of Interlocked Structures. Vol. 171, pp. 1–75.
Takeuchi, D. see Osakada, K.: Vol. 171, pp. 137–194.
Tan, K. L. see Kang, E. T.: Vol. 106, pp. 135–190.
Tanaka, H. and *Shibayama, M.*: Phase Transition and Related Phenomena of Polymer Gels. Vol. 109, pp. 1–62.
Tanaka, T. see Penelle, J.: Vol. 102, pp. 73–104.
Tauer, K. see Guyot, A.: Vol. 111, pp. 43–66.
Tendler, S. J. B. see Ellis, J. S.: Vol. 193, pp. 123–172.
Tenhu, H. see Aseyev, V. O.: Vol. 196, pp. 1–86.
Teramoto, A. see Sato, T.: Vol. 126, pp. 85–162.
Terent'eva, J. P. and *Fridman, M. L.*: Compositions Based on Aminoresins. Vol. 101, pp. 29–64.
Terry, A. E. see Rastogi, S.: Vol. 180, pp. 161–194.
Theodorou, D. N. see Dodd, L. R.: Vol. 116, pp. 249–282.
Thomson, R. C., Wake, M. C., Yaszemski, M. J. and *Mikos, A. G.*: Biodegradable Polymer Scaffolds to Regenerate Organs. Vol. 122, pp. 245–274.
Thünemann, A. F., Müller, M., Dautzenberg, H., Joanny, J.-F. and *Löwen, H.*: Polyelectrolyte complexes. Vol. 166, pp. 113–171.
Tieke, B. see v. Klitzing, R.: Vol. 165, pp. 177–210.
Tobita, H. see Nomura, M.: Vol. 175, pp. 1–128.
Tokita, M.: Friction Between Polymer Networks of Gels and Solvent. Vol. 110, pp. 27–48.
Traser, S. see Bohrisch, J.: Vol. 165, pp. 1–41.
Tries, V. see Baschnagel, J.: Vol. 152, p. 41–156.
Trimmel, G., Riegler, S., Fuchs, G., Slugovc, C. and *Stelzer, F.*: Liquid Crystalline Polymers by Metathesis Polymerization. Vol. 176, pp. 43–87.
Tsujii, Y., Ohno, K., Yamamoto, S., Goto, A. and *Fukuda, T.*: Structure and Properties of High-Density Polymer Brushes Prepared by Surface-Initiated Living Radical Polymerization. Vol. 197, pp. 1–47.
Tsuruta, T.: Contemporary Topics in Polymeric Materials for Biomedical Applications. Vol. 126, pp. 1–52.

Uemura, T., Naka, K. and *Chujo, Y.*: Functional Macromolecules with Electron-Donating Dithiafulvene Unit. Vol. 167, pp. 81–106.
Ungar, G., Putra, E. G. R., de Silva, D. S. M., Shcherbina, M. A. and *Waddon, A. J.*: The Effect of Self-Poisoning on Crystal Morphology and Growth Rates. Vol. 180, pp. 45–87.
Usov, D. see Rühe, J.: Vol. 165, pp. 79–150.
Usuki, A., Hasegawa, N. and *Kato, M.*: Polymer-Clay Nanocomposites. Vol. 179, pp. 135–195.
Uyama, H. and *Kobayashi, S.*: Enzymatic Synthesis and Properties of Polymers from Polyphenols. Vol. 194, pp. 51–67.
Uyama, H. and *Kobayashi, S.*: Enzymatic Synthesis of Polyesters via Polycondensation. Vol. 194, pp. 133–158.
Uyama, H. see Kobayashi, S.: Vol. 121, pp. 1–30.
Uyama, Y.: Surface Modification of Polymers by Grafting. Vol. 137, pp. 1–40.

Vancso, G. J., Hillborg, H. and *Schönherr, H.*: Chemical Composition of Polymer Surfaces Imaged by Atomic Force Microscopy and Complementary Approaches. Vol. 182, pp. 55–129.
Varma, I. K. see Albertsson, A.-C.: Vol. 157, pp. 99–138.
Vasilevskaya, V. see Khokhlov, A.: Vol. 109, pp. 121–172.

Vaskova, V. see Hunkeler, D.: Vol. 112, pp. 115–134.
Verdugo, P.: Polymer Gel Phase Transition in Condensation-Decondensation of Secretory Products. Vol. 110, pp. 145–156.
Veronese, F. M. see Pasut, G.: Vol. 192, pp. 95–134.
Vettegren, V. I. see Bronnikov, S. V.: Vol. 125, pp. 103–146.
Vilgis, T. A. see Holm, C.: Vol. 166, pp. 67–111.
Viovy, J.-L. and *Lesec, J.*: Separation of Macromolecules in Gels: Permeation Chromatography and Electrophoresis. Vol. 114, pp. 1–42.
Vlahos, C. see Hadjichristidis, N.: Vol. 142, pp. 71–128.
Voigt, I. see Spange, S.: Vol. 165, pp. 43–78.
Volk, N., Vollmer, D., Schmidt, M., Oppermann, W. and *Huber, K.*: Conformation and Phase Diagrams of Flexible Polyelectrolytes. Vol. 166, pp. 29–65.
Volksen, W.: Condensation Polyimides: Synthesis, Solution Behavior, and Imidization Characteristics. Vol. 117, pp. 111–164.
Volksen, W. see Hedrick, J. L.: Vol. 141, pp. 1–44.
Volksen, W. see Hedrick, J. L.: Vol. 147, pp. 61–112.
Vollmer, D. see Volk, N.: Vol. 166, pp. 29–65.
Voskerician, G. and *Weder, C.*: Electronic Properties of PAEs. Vol. 177, pp. 209–248.

Waddon, A. J. see Ungar, G.: Vol. 180, pp. 45–87.
Wagener, K. B. see Baughman, T. W.: Vol. 176, pp. 1–42.
Wagner, E. and *Kloeckner, J.*: Gene Delivery Using Polymer Therapeutics. Vol. 192, pp. 135–173.
Wake, M. C. see Thomson, R. C.: Vol. 122, pp. 245–274.
Wandrey, C., Hernández-Barajas, J. and *Hunkeler, D.*: Diallyldimethylammonium Chloride and its Polymers. Vol. 145, pp. 123–182.
Wang, K. L. see Cussler, E. L.: Vol. 110, pp. 67–80.
Wang, S.-Q.: Molecular Transitions and Dynamics at Polymer/Wall Interfaces: Origins of Flow Instabilities and Wall Slip. Vol. 138, pp. 227–276.
Wang, S.-Q. see Bhargava, R.: Vol. 163, pp. 137–191.
Wang, T. G. see Prokop, A.: Vol. 136, pp. 1–52; 53–74.
Wang, X. see Lin, T.-C.: Vol. 161, pp. 157–193.
Watanabe, K. see Hikosaka, M.: Vol. 191, pp. 137–186.
Webster, O. W.: Group Transfer Polymerization: Mechanism and Comparison with Other Methods of Controlled Polymerization of Acrylic Monomers. Vol. 167, pp. 1–34.
Weder, C. see Voskerician, G.: Vol. 177, pp. 209–248.
Weis, J.-J. and *Levesque, D.*: Simple Dipolar Fluids as Generic Models for Soft Matter. Vol. 185, pp. 163–225.
Whitesell, R. R. see Prokop, A.: Vol. 136, pp. 53–74.
Williams, R. A. see Geil, P. H.: Vol. 180, pp. 89–159.
Williams, R. J. J., Rozenberg, B. A. and *Pascault, J.-P.*: Reaction Induced Phase Separation in Modified Thermosetting Polymers. Vol. 128, pp. 95–156.
Winkler, R. G. see Holm, C.: Vol. 166, pp. 67–111.
Winnik, F. M. see Aseyev, V. O.: Vol. 196, pp. 1–86.
Winter, H. H. and *Mours, M.*: Rheology of Polymers Near Liquid-Solid Transitions. Vol. 134, pp. 165–234.
Wittmeyer, P. see Bohrisch, J.: Vol. 165, pp. 1–41.
Wood-Adams, P. M. see Anantawaraskul, S.: Vol. 182, pp. 1–54.
Wu, C.: Laser Light Scattering Characterization of Special Intractable Macromolecules in Solution. Vol. 137, pp. 103–134.

Wu, C. see *Zhang, G.*: Vol. 195, pp. 101–176.
Wunderlich, B. see *Sumpter, B. G.*: Vol. 116, pp. 27–72.

Xiang, M. see *Jiang, M.*: Vol. 146, pp. 121–194.
Xie, T. Y. see *Hunkeler, D.*: Vol. 112, pp. 115–134.
Xu, P., Singh, A. and *Kaplan, D. L.*: Enzymatic Catalysis in the Synthesis of Polyanilines and Derivatives of Polyanilines. Vol. 194, pp. 69–94.
Xu, P. see *Geil, P. H.*: Vol. 180, pp. 89–159.
Xu, Z., Hadjichristidis, N., Fetters, L. J. and *Mays, J. W.*: Structure/Chain-Flexibility Relationships of Polymers. Vol. 120, pp. 1–50.

Yagci, Y. and *Endo, T.*: N-Benzyl and N-Alkoxy Pyridium Salts as Thermal and Photochemical Initiators for Cationic Polymerization. Vol. 127, pp. 59–86.
Yamaguchi, I. see *Yamamoto, T.*: Vol. 177, pp. 181–208.
Yamamoto, T.: Molecular Dynamics Modeling of the Crystal-Melt Interfaces and the Growth of Chain Folded Lamellae. Vol. 191, pp. 37–85.
Yamamoto, T., Yamaguchi, I. and *Yasuda, T.*: PAEs with Heteroaromatic Rings. Vol. 177, pp. 181–208.
Yamamoto, S. see *Tsujii, Y.*: Vol. 197, pp. 1–47.
Yamaoka, H.: Polymer Materials for Fusion Reactors. Vol. 105, pp. 117–144.
Yamazaki, S. see *Hikosaka, M.*: Vol. 191, pp. 137–186.
Yannas, I. V.: Tissue Regeneration Templates Based on Collagen-Glycosaminoglycan Copolymers. Vol. 122, pp. 219–244.
Yang, J. see *Geil, P. H.*: Vol. 180, pp. 89–159.
Yang, J. S. see *Jo, W. H.*: Vol. 156, pp. 1–52.
Yasuda, H. and *Ihara, E.*: Rare Earth Metal-Initiated Living Polymerizations of Polar and Nonpolar Monomers. Vol. 133, pp. 53–102.
Yasuda, T. see *Yamamoto, T.*: Vol. 177, pp. 181–208.
Yaszemski, M. J. see *Thomson, R. C.*: Vol. 122, pp. 245–274.
Yoo, T. see *Quirk, R. P.*: Vol. 153, pp. 67–162.
Yoon, D. Y. see *Hedrick, J. L.*: Vol. 141, pp. 1–44.
Yoshida, H. and *Ichikawa, T.*: Electron Spin Studies of Free Radicals in Irradiated Polymers. Vol. 105, pp. 3–36.

Zhang, G. and *Wu, C.*: Folding and Formation of Mesoglobules in Dilute Copolymer Solutions. Vol. 195, pp. 101–176.
Zhang, H. see *Rühe, J.*: Vol. 165, pp. 79–150.
Zhang, Y.: Synchrotron Radiation Direct Photo Etching of Polymers. Vol. 168, pp. 291–340.
Zheng, J. and *Swager, T. M.*: Poly(arylene ethynylene)s in Chemosensing and Biosensing. Vol. 177, pp. 151–177.
Zhou, H. see *Jiang, M.*: Vol. 146, pp. 121–194.
Zhou, Z. see *Abe, A.*: Vol. 181, pp. 121–152.
Zubov, V. P., Ivanov, A. E. and *Saburov, V. V.*: Polymer-Coated Adsorbents for the Separation of Biopolymers and Particles. Vol. 104, pp. 135–176.

Subject Index

AAm *I* 19
Addition polymerization *I* 119
AIBN *I* 50
Alkyne *I* 137
Aluminum oxide *I* 34
Anchored polymers, gradients *II* 86
ATRP *I* 6, 50, 110; *II* 3, 63, 73, 92, 125, 129
–, surface-initiated *I* 8
–, –, SiP *I* 33
AuNPs *I* 34, 128, 142
–, PMMA-coated *I* 36

Benzophenone *I* 50
Bio-fouling *I* 37
Biointerface *I* 37
Biomedical applications, photoiniferters *I* 67
Biomimetic surface *I* 93
BMPUS *II* 81
Brushes *I* 1; *II* 152
–, charged *II* 149, 170
–, dry *I* 24
–, high-density, applications *I* 33
–, mixed *I* 61
–, 'nonfouling' *I* 39
–, osmotic *II* 80, 85
–, photoinitiated synthesis *I* 47
–, salted *II* 79, 84
–, swollen *I* 17

Carbanions *I* 111
Carbenes, N-heterocyclic *I* 163
Carbocations *I* 111
Carbon black surface *I* 34, 126
Carbon nanotubes *I* 167
Carboxamides *II* 34
CdS/SiO$_2$ nanoparticles *I* 34
Cell adhesion *I* 68
Charge regulation *I* 30
Charged polymers *II* 149, 156, 170
Chemical force microscopy (CFM) *II* 67

Chitosan *II* 45
Chloromethylstyrene, branching *I* 80
CMPE-SAMs *II* 71
Collapse *II* 154
Contact angle, wettability *II* 68
Corona treatment *II* 59
Crowding *I* 7
Cyclodextrin *II* 14
Cytokines *I* 104

DCA (dynamic contact angle) *II* 68
Debye-Hückel *II* 158
Decanethiol monolayer *II* 42
Dendrimer *II* 1
– analogs *II* 43
DEPN *I* 15
Diblock copolymer brushes *II* 125, 179
Dicyclopentadiene *II* 158
Diethyl diallylmalonate *I* 153
Dip coating *I* 49
1,1-Diphenylethylene *I* 116
Dithiocarbamates, photolysis *I* 68, 70

Electrostriction *I* 27
End-functionalized polymers *II* 150
Epidermal growth factor/polystyrene *II* 64
Epoxy silane *II* 61, 129

Films, ultra-thin *I* 24
Free radical polymerization *I* 4
Freely-jointed chain model *II* 153

Gantrez *II* 30
Glass transition, polymer film *I* 25
Glycidyl methacrylate GMA) *I* 30
Gold, photoinitiated grafting *I* 57
Gold films, PAA *II* 5
Gold nanoparticles *I* 34, 128, 142
– –, SAM *II* 105
Gold surfaces *I* 124, 142

Gouy-Chapman length *II* 174
Gradient-graft copolymers *II* 65
Gradients, orthogonal *II* 52, 111
Graft-on-a-graft *II* 1
Graft polymerization *I* 1
– –, livingness *I* 74
Grafting, photoinitiated *I* 47
Grafting-to-/from *I* 49, 110, 137, 142; *II* 54, 60, 63, 126, 151
Grafts, hyperbranched *I* 82
Graphite/polystyrene *I* 128
Grubbs catalysts *I* 141, 157

Heptadecafluorodecyl acrylate (HFA) *II* 134

Implant rejection *I* 68
Iniferters *I* 47, 50, 69; *II* 3
Iodine *I* 6

Living polymerization *I* 111
Living radical polymerization, brushes *I* 1, 5

Macroinitiator *II* 66
Macrophages *I* 103
Magnetic nanoparticles *I* 34
Melamine dendrimers *II* 45
Mercaptoundecanoic acid, self-assembled monolayer *II* 5
Merrifield-type resins *I* 155
Metathesis polymerization *I* 137
Methacrylates, monoliths *I* 157
Mica *I* 36
Microfluidics *I* 47
MMA *I* 11, 33, 116
$Mn_2(CO)_{10}$ *II* 24
Molybdenum *I* 140
Monoliths *I* 156
–, ROMP-derived *I* 158
Multigeneration hyperbranched graft *I* 84
Mushroom conformation *I* 3; *II* 56, 152
Mushroom-to-brush crossover/transition *I* 19; *II* 52

Nanocomposites, hyperbranched *II* 30
Nanoparticles, gold *I* 34
–, gradients, orthogonal *II* 111
–, magnetic *I* 34
NIPAA *II* 127

Nitroxides *I* 6
NMP, surface-initiated *I* 15, 32
Norbornadiene *I* 139, 158

OEGMA *I* 13
Oligo(ethylene glycol) methyl methacrylate *I* 13
Oligothiophene *II* 26
Organic/inorganic hybrid *II* 39
Orthogonal gradients *II* 52, 111
Osmotic brush *II* 80, 85
2-Oxazolines *I* 129

P2VP *II* 62, 128
P4VP *I* 20
PAA *II* 70
–, dry thickness *II* 90
–, graft-on-a-graft *II* 4
–, hyperbranched grafts *II* 1, 4
–, solution properties *II* 83
PAA/Au *II* 9
PAA/P2VP *II* 62
PAAm *I* 10, 20; *II* 70, 73, 106
–, grafting density *II* 73
PAMAM *II* 14, 21, 30, 43
Patterned hyperbranched grafts *II* 17
PBd-b-PS *I* 124
Pd(0) catalysts *II* 28
Pd(II)-coordinated polymers, dendritic hyperbranched grafts *II* 41
PDMA *I* 15, 21; *II* 132
PEGMA *II* 60, 128
PEMA *I* 30
Pentafluorostyrene *II* 134
PEO *I* 38
PHEMA *II* 70, 114
–, gradient *II* 92
2-Phenylprop-2-yl dithiobenzoate *I* 15
Phosphatidylcholin *I* 68
Photoiniferter *I* 50, 68
Photoinitiated polymerization (SIP) *I* 49
Photosensitizers *I* 50
PIB *I* 131
PIBVE *II* 45
Pincer compounds II 39-41
Pincus brush height *II* 174
PMAA *I* 29
PMMA brushes *I* 8, 16, 17; *II* 70, 88
PMMA/PS *II* 130
PNIPAM *I* 21; *II* 22, 23

Subject Index

Poly(acrylamide) *I* 68; *II* 127
Poly(acrylic acid) *II* 128
–, hyperbranched grafts *II* 1, 4
Poly(allyl alcohol)s *II* 24
Poly(amidoamidoamine) *II* 14, 21, 30, 43
Poly(N,N-dimethylacrylamide) *I* 15
Poly(dimethyl aminomethyl methacrylate)
 PDMAEMA *II* 96, 111, 130
Poly(etheylene glycol) *I* 68
Poly(ethylene glycol) *I* 38; *II* 60
– monomethacrylate (PEGMA) *II* 128
Poly(ethylene oxide) *II* 114, 151
–, comb-like *II* 60
Poly(ethyleneimine), hyperbranched grafts *II* 37
Poly(glycidyl methacrylate) *II* 61, 129
Poly(heptadecafluorodecyl acrylate) (PHFA) *II* 143
Poly(isobutyl vinyl ether) *II* 45
Poly(N-isopropylacrylamide) *I* 21; *II* 22, 23, 127
Poly(maleic anhydride) *II* 30, 34
Poly(methacrylic acid) *I* 29
Poly(methyl acrylate) *II* 130
Poly(pentafluoro styrene) (PPFS) *II* 143
Poly(pentafluoropropyl acrylate) (PPFA) *II* 143
Poly(sodium acrylate) *II* 14
Poly(styrene-b-$tert$-butyl acrylate) *II* 129
Poly(styrene-co-divinylbenzene), monoliths *I* 157
Poly($tert$-butyl acrylate) *II* 62
Poly(trifluoroethyl acrylate) (PTFA) *II* 143
Poly(2-vinylpyridine) (P2VP) *II* 128
Poly(4-vinylpyridine) *I* 20
Polyampholytes *II* 95
Polyanhydride *II* 1
Polyelectrolytes *II* 149, 156
–, surface-bound *II* 78
Polyesters *I* 126
Polyethylene *II* 171
–, corona discharge *II* 59
–, hyperbranched grafts *II* 1, 19, 27
–, scaling diagram *II* 175
Polyglycidol, hyperbranched grafts *II* 36
Polyisobutylene *I* 131
Polyisoprene *II* 128
Polymer brushes *I* 1
Polymer end-points *II* 160

Polymer gradients *II* 59
Polypropylene, hyperbranched grafts *II* 19, 26
Polysiloxane, hyperbranched grafts *II* 1, 39
Polystyrene *I* 15, 120, 131; *II* 129, 151
PPEI *I* 130
1-Propoxyethyl methacrylate *I* 30
Protein adhesion *II* 57
Protein adsorption *I* 37, 68; *II* 52, 60, 114
Protein/brush surface *I* 37
PS-b-PI *I* 124
PS-PEO *II* 151
PS-PVP *II* 151
PTBA *II* 5

Quartz crystal microbalance *I* 74

RAFT *I* 6
–, surface-initiated *I* 15
RATPR *II* 131
Reversible addition fragmentation transfer (RAFT) *II* 132, 135
Rigid rod *I* 157
ROMP *I* 137; *II* 36

Salted brush regime *II* 79, 84
Scaling theory *II* 149
Schrock catalysts *I* 140, 157
SCS-Pd(II) *II* 41
Self-assembled monolayer *I* 10, 48, 107, 131, 144; *II* 5, 54, 105, 126
Self-condensed vinyl polymerization *II* 3
Self-consistent field theory *II* 149
Si(111) *I* 143
Si/SiO$_2$//PS-b-PMMA *II* 137
Silane-coupling agent *I* 8
Siloxanes *II* 39
SiO$_x$, grafting *I* 53, 124
Solvents *II* 155, 169
Spin casting *I* 49
Stimuli-responsive films *II* 125
Stretching factor *II* 167
Styrene *I* 15
– / MMA *II* 129
Surface-grafted polymer assemblies, gradients *II* 71
Surface-initiated polymerization *I* 6, 47, 107, 110
Swelling *II* 154

Tellanyls *I* 6
TEMPO *I* 15, 110
Tethering polymer *I* 1, 7
Thin films *II* 125
2-Thiopheneethyleneamine *II* 25

N-Vinyl-2-pyrrolidone *I* 130
2-Vinylthiophene *I* 127

Wet thickness, PAA *II* 83
Wettability *II* 68

Printing: Krips bv, Meppel
Binding: Stürtz, Würzburg